Environmental Populism

Mark Beeson

Environmental Populism

The Politics of Survival in the Anthropocene

Mark Beeson
School of Social Sciences
University of Western Australia
Perth, WA, Australia

ISBN 978-981-13-7476-0   ISBN 978-981-13-7477-7 (eBook)
https://doi.org/10.1007/978-981-13-7477-7

© The Editor(s) (if applicable) and The Author(s), under exclusive licence to Springer Nature Singapore Pte Ltd. 2019
This work is subject to copyright. All rights are solely and exclusively licensed by the Publisher, whether the whole or part of the material is concerned, specifically the rights of translation, reprinting, reuse of illustrations, recitation, broadcasting, reproduction on microfilms or in any other physical way, and transmission or information storage and retrieval, electronic adaptation, computer software, or by similar or dissimilar methodology now known or hereafter developed.
The use of general descriptive names, registered names, trademarks, service marks, etc. in this publication does not imply, even in the absence of a specific statement, that such names are exempt from the relevant protective laws and regulations and therefore free for general use.
The publisher, the authors and the editors are safe to assume that the advice and information in this book are believed to be true and accurate at the date of publication. Neither the publisher nor the authors or the editors give a warranty, express or implied, with respect to the material contained herein or for any errors or omissions that may have been made. The publisher remains neutral with regard to jurisdictional claims in published maps and institutional affiliations.

Cover Image: © Tobias Titz / Getty Images

This Palgrave Macmillan imprint is published by the registered company Springer Nature Singapore Pte Ltd.
The registered company address is: 152 Beach Road, #21-01/04 Gateway East, Singapore 189721, Singapore

For Greta Thunberg and her generation

Acknowledgments

A number of friends and colleagues have read, commented on or otherwise contributed to this book. Although none of them should be held responsible for the contents, I thank Bruce Campbell, Rob Cover, Ann Firth, Kanishka Jayasuriya, Matt McDonald and Ben Reilly for their help.

Contents

1 **The State of the World**   1
  What Do We Know About Climate Change?   3
  The Social Construction of Reality   4
  Just the Facts, Ma'am   8
  The End of Civilization as We Know It?   10
  Conclusion   14

2 **The Paradoxes of Development**   21
  It's Just Not Fair   22
  Can We Cope with Capitalism?   24
  A Planned Alternative?   28
  Are Other Pathways Possible?   32
  Conclusion   34

3 **The Limits to International Cooperation**   41
  Making Sense of International Cooperation   42
  States: Part of the Solution?   44
  Why Is International Cooperation So Difficult?   46
  Coming Up with a Common Agenda   48
  Scaling Down?   50
  Conclusion   54

4 Debating Populism 63
The Death of Democracy? 64
Populism in Theory and Practice 67
We, the People 69
The Individual Versus the Collective 72
The Way We Were (and Could Be)? 74
Social Movements and Global Civil Society 76
Conclusion 79

5 Populism in Practice 87
(Mostly) Bad Populism 88
The Political Economy of Populism 90
The Contradictions of Populism 93
The Migration Crisis 95
Lamentations of the Lucky Country 100
Conclusion 103

6 Conclusion: The Unsustainable Status Quo 111
What Is to Be Done? 113
Australia Saves the World… 115
Hope Springs Eternal 117

Index 123

Acronyms

AFD	Alternative for Germany
ALP	Australian Labor Party
BRI	Belt and Road Initiative
CCP	Chinese Communist Party
CNAS	Center for New American Security
CSIS	Center for Strategic and International Studies
EPA	Environmental Protection Agency
EU	European Union
GFC	Global financial crisis
GONGO	Government-organized NGOs
IFIs	International financial institutions
IGO	Intergovernmental organization
IPCC	Intergovernmental Panel on Climate Change
IR	International relations
NGOs	Nongovernment organizations
OWS	Occupy Wall Street
PRC	People's Republic of China
UNEP	United Nations Environment Programme
UNFCCC	UN Framework Convention on Climate Change

Introduction

This is not the sort of book academics usually produce, nor is it intended to be. Some prospective readers may be breathing a sigh of relief at this point, but I feel obliged to warn the more 'scholarly' types that they may be disappointed. To be sure, there are many references to, and discussions on, the relevant academic literature, but the book is intended for a general audience and is unapologetically polemical, subjective and tendentious. By the end of it I hope to convince you, dear reader, that it is not only possible to take action on climate change, but it is up to all of us to compel our political and economic leaders to get on with it before it's too late. The best mechanism for achieving this is an unlikely one, perhaps, but one that could conceivably generate the sort of social and political momentum that seems necessary if real change in the way we live is to occur.

Thus far, populism has generally had a deservedly bad press. It is, after all, usually associated with odious megalomaniacs and despots of one sort or another, who claim to be the embodiment of the popular will. Chapters 4 and 5 consider the nature of populism in detail, including its generally unfortunate history. There are, however, some—rather encouraging—historical examples of populism that have championed 'progressive' causes, often in response to major crises of the sort that seem to have become an inescapable feature of contemporary life. Indeed, it is remarkable how disorderly the international order has actually become, and how incapable the current crop of world leaders seems to be of doing anything about it.

To be fair, some of the challenges we currently face are unprecedented, and it is far from certain whether the likes of Bismarck, Churchill, Roosevelt (either of them), Monnet or Schumann would have been any more capable of addressing them than the present generation. Some readers will no doubt be astonished to see the names of the founding fathers of what came to be the European Union (EU) included in that list of historical luminaries. I should confess at the outset that I still regard the EU as the most important political experiment in the history of the world bar none. I say more about this in Chap. 3, and about all of the problems that confront those of us interested in trying to develop sustainable forms of political cooperation, which seem a necessary precondition for the environmental variety.

Chapter 1 spells out the nature and extent of the environmental problems that are presenting such a fundamental and possibly irresolvable challenge for our political and economic elites, and why this unprecedented era merits its own name: the Anthropocene. Indeed, Chap. 2 explains why dealing with climate change is not simply a political or technological issue—although they are formidable enough—but it is also a profound economic challenge. Throughout the book, I illustrate the issues primarily by looking at the US and China—because they are some of the biggest contributors to the problems and pivotal to any possible solutions—and on Australia, because that's where I live and where I can act.

It's not necessary to be an unreconstructed Marxist—although it probably helps—to recognize that there's a potentially fundamental tension between finite planetary boundaries and an inherently expansionary, consumption-oriented capitalist economy. There's also the perennial problem of continuing economic inequality, which places a major hurdle in the way of possible collective action. The nature of, and obstacles to, collective action is another of the recurrent themes that feature throughout the book. Indeed, the potentially good news in this regard is that we are unambiguously in this together, and more people are beginning to recognize that—even if they are not always in a position to do much about it.

Never before in human history has there been a problem with quite such a global impact. True, some parts of the world are likely to be much more badly affected than others in the short term, but no place is immune. The bad news is that those who have done least to cause the problem of global warming are likely to suffer most from its impact and may not be well placed to respond to the challenge. Recognizing that this is not simply a practical problem, but one with complex ethical implications that we shall all have to think about is an important first step. The solutions to such problems—if, indeed, there are any—will not be easily developed or implemented. The immediate task, however, is to compel policymakers around the world—whether they are democrats, autocrats or something in between—to act.

You and I can also do our bit, and not just in the form of recycling, using public transport or going vegan. Laudable as such things may be, skeptics are right to point out that our individual efforts will not address the underlying structural factors that allow—even encourage—certain forms of behavior and activity to occur. For all their manifest problems and shortcomings, at this moment in human history, only states can actually do that. To be sure, there are plenty of non-state actors undertaking worthy and even inspiring activities, but states largely determine what goes on within the places we inhabit, no matter how artificial, arbitrary and accidental those political boundaries may be.

That is why I suggest that the place 'we'—in this case you and I, but ultimately humanity in general—need to start is where we live. Collective action starts at home, literally and metaphorically. For those of us fortunate to live in democracies, political participation is comparatively easy and generally not a matter of life and death. It is, therefore, up to citizens of democracies to lead that way and take action. Not because we are better, but because we can. If the leaders of the world's democracies really make a serious effort to address climate change, perhaps even one that involves making national sacrifices and contributions to other less fortunate states, it really could make a difference. If nothing else, the very act of participating and feeling slightly more empowered and virtuous about doing so might be good for our collective mental health. Environmental populism might not be quite as mad as it sounds.

CHAPTER 1

The State of the World

Much of the content of this chapter may be—or should be—familiar to readers of a book such as this. Indeed, there is inevitably an element of 'preaching to the converted' with these sorts of volumes. Some potential readers are either sick of hearing about the world's environmental problems, already have strong opinions about their possible significance, or subscribe to views that are unlikely to be influenced by, or open to, an objective analysis of 'the facts'. More to the point, they are unlikely to read this sort of work in the first place. This is of course part of the problem facing those of us who feel that something ought to be done, and that it will necessarily involve large-scale collective action. If large parts of the global population either have stopped listening or never knew there was something important to listen to in the first place, then the chances for meaningful collective action are not good.

The intention of this chapter is not to give an exhaustive account of the science of climate change, but to outline briefly some of the—sadly, all-too-plausible—evidence that has been accumulated about the causes and consequences of climate change. As I shall explain, more than just our collective understanding of the facts is in play here. But if we can't at least agree on some of the basic parameters within which the following discussion is to take place, it's difficult to know quite where to start. No doubt, the following narrative is colored by my own prejudices, biases and

© The Author(s) 2019
M. Beeson, *Environmental Populism*,
https://doi.org/10.1007/978-981-13-7477-7_1

beliefs—especially about the very urgent need for unprecedented and still unlikely action. But I have no particular ideological or theological axe to grind, just an overwhelming sense of foreboding that time is running out if we are to save ourselves and bequeath something that resembles a life worth living to future generations.

The difficulties that confront human beings from taking collective action will be one of the recurring themes of this book. It is not controversial to suggest why this is important, even if there is widespread disagreement about whether it may be achievable on the sort of planetary scale that seems necessary. Cooperative social action is necessary to provide collective goods or services such as law and order, defense, public welfare or anything else that individuals acting alone cannot accomplish. Such outcomes have usually been organized by states in the modern period, and there are major questions about whether the provision of environmental collective goods can actually be scaled up to the international level, as we shall see in Chap. 3.

Suffice to say at this point that some of the leading lights in social theory have come to very different conclusions about whether human beings are capable of acting collectively in the first place. Mancur Olson, for example, famously argued that 'unless there is coercion or some other special device to make individuals act in their common interest, *rational, self-interested individuals will not act to achieve their common or group interests*'.[1] While Olson's highly influential reading of human behavior undoubtedly captures something important about the tension between individuals and their willingness to cooperate, one of the real problems is, what happens if such insights are taken to be an accurate reflection of unchanging 'human nature'. Such assumptions have become foundational in mainstream economics and international relations (IR) theory. As another social science luminary, Elinor Ostrom, points out though, 'what makes these models so dangerous—when they are used metaphorically as the foundation for policy—is that the constraints assumed to be fixed for the purpose of analysis are taken on faith as being fixed in empirical settings, unless external authorities change them'.[2] As we shall see, in the current international order there simply aren't any 'external authorities' with the capability to change things to ensure international cooperation—even if we can agree on what that cooperation ought to look like and achieve. This makes agreement on the facts doubly important.

WHAT DO WE KNOW ABOUT CLIMATE CHANGE?

There has never been a problem like climate change. No other problem—with the possible exception of all-out nuclear war between two or more of the great powers—threatens the continuing existence of the human race in quite the way that climate change seems to. I say 'seems to', because one of the distinguishing features of the climate change debate is that there is still a good deal of uncertainty about some of the underlying 'facts', especially about the long-term impact and extent of global warming.[3] It could hardly be otherwise, given that the challenge presented by global warming in particular is unprecedented, and climate scientists must use their judgment and best guesses about when and how various processes will unfold and what their impact will be. Disagreement about the nature of any policy challenge is a problem at any time; in an era of so-called fake news, it is a major impediment to meaningful action of any sort.[4] It is a potentially fatal problem as far as large-scale collective action is concerned.

The comparison with the threat posed by nuclear war highlights a couple of other features of the overall environmental debate that are worth emphasizing at the outset. The unprecedented nature of the problem and the necessary uncertainty it generates help to explain why climate change is such a 'wicked problem' and resistant to action, despite widespread agreement within the relevant scientific community about its possible causes and, more importantly at this point of the discussion, its possible impact.[5] By contrast, there is little disagreement about the likely impact of nuclear weapons.[6] We have two salutatory and illuminating historical examples of the devastation even relatively small nuclear devices can cause. While there may be a bit of doubt about the number of people who would be killed by specific weapons in particular areas, no one doubts their individual and collective impact would be apocalyptic, especially if it proved difficult to stop a rapid escalation of any initial conflict. Nuclear weapons have become unbelievably powerful and 'efficient' in the 70 years or so since Hiroshima and Nagasaki were obliterated.

As a result, (almost) no one is arguing that large-scale nuclear war would be anything other than potentially catastrophic. Consequently, most people seem to think it's worth making an effort to avoid nuclear war. Climate change, by contrast—to use a possibly unfortunate metaphor—is something of a slow burner. True, climate scientists continually point to the fact that time appears to be running out as far as meaningful action is concerned.[7] Even so, we're still talking about years, possibly decades, before the impacts of climate change are likely to wreak havoc on some of the wealthier parts of the world where much of today's decision-making processes take place. In the meantime, much fiddling—and debating—can take place while the planet heats up, and those least responsible for causing the problems suffer the most. The uneven impact of climate change is another of the problems that makes action particularly difficult, not least because the most vulnerable are often the least powerful and influential.[8]

One final initial comparative point is worth making about climate change and the possibly apocalyptic threat posed by military conflict between the major powers. Few people among influential policy elites around the world, especially in the military establishments that dominate security debates, think that spending large amounts of money on weapons of mass destruction is a bad idea.[9] If a state's potential rivals and enemies have them, the argument goes, it would be irresponsible not to do the same—even to the point of 'mutually assured destruction'.[10] This is not the place to dissect the logic of nuclear warfare and deterrence, but it is important to recognize that some ideas about reality—especially the strategic and geopolitical sort—carry more weight than others. The slightly optimism-inducing aspect of this point is that even military elites and strategic thinkers now seem to recognize that climate change is a growing strategic threat and therefore ought to be taken seriously.[11] Before considering the alarming conclusions strategic thinkers predictably draw from this (see Chap. 3), it is important to say something about the facts as far as we understand them. This is not as straightforward as it sounds, and the conclusions that are drawn from them—even by those who accept the proposition that indisputable facts actually exist anymore—are not necessarily objective.

The Social Construction of Reality

Let me confess at the outset that I am not a 'climate change expert'. If the world had to wait for me to figure out what the 'greenhouse effect' was, we would have been waiting for an awfully long time. My area of expertise, such as it is, is international politics. I like to think I know something about this, although not everyone else working in the same field would necessarily agree. Indeed, many of my colleagues disagree with the conclusions I come to about my chosen subject, a sentiment this book is unlikely to change, I fear. But I am not emphasizing this point in order to win over the reader with an endearing display of modesty and self-deprecation; my intention is to highlight a fundamental difference between the social and the natural sciences, and—despite my rather unfashionable admiration for the latter—the limitations of both.[12]

For the purposes of this discussion, there are two important points to make about the way we understand the world. First, it is inevitably a collective effort undertaken by 'us'.[13] True, some people—the likes of Isaac Newton and Albert Einstein, for example—make a bigger contribution than most, but the production of knowledge is ultimately a social process. These days there's a lot more knowledge being produced, even if the quality varies. Some say the last person who knew just about everything interesting about the time he lived in was Max Weber, the German sociologist, who was born in 1864. Whatever the merits of this claim, it captures something important about the sheer volume of stuff there is to know these days. Consequently, an intellectual division of labor exists in which people specialize in increasingly narrow areas of expertise. As a result, when something like 97 percent of people who describe themselves as climate change scientists agree on the underlying dynamics of global warming,[14] for example, I'm happy to take their word for it.

By contrast, the social sciences are by their very nature much more contested and uncertain.[15] Any discipline that takes human beings and their individual and collective actions as its central focus of attention could hardly be otherwise. Human beings are a diverse bunch and their behavior continues to display remarkable degrees of variation despite much excited talk about the possible impact of 'globalization'.[16] Significantly, claims about the inevitability of a possible process of 'convergence' on one developmental template or another have rather notably gone out of fashion.[17] On the contrary, questions of identity and difference are becoming more important rather than less.[18] There is no unambiguously right way of living, even if 'we' might agree that some look more desirable and sustainable than others. As we shall see, this is potentially something of a problem for action on climate change: if people in different nations, social groups, religious communities, classes or regions have fundamentally different views about the nature of reality, public policy priorities and much else, collective action is going to be difficult. The capacity for 'learning' that informs so much of the literature about policymaking looks dubious, especially outside the rarified enclaves of what currently passes for global governance.[19]

In this regard, what ought to be a collective strength—that is, our ability to understand complex environmental processes and develop possible policy responses—is increasingly being contested, and even seen by some as a source of something akin to ideological or ideational imperialism.[20] Although what we think of as modern science may not have originated exclusively in the West,[21] it has come to be associated with a particular way of thinking and acting that is overwhelmingly associated with the Enlightenment in Europe.[22] At the center of the scientific paradigm that became increasingly influential in the sixteenth and seventeenth centuries was the idea that universal laws of nature could be 'discovered' and that they might be applied in the same way across the world. The quintessential example of this possibility, perhaps, was the realization by Copernicus that the Earth revolved around the Sun rather than vice versa. Despite the fact that there are not many people who disagree with this remarkable *and accurate* insight these days, the reception of such claims by the 'power elite' of the day was very different.[23]

Unfortunately, a hostile reception to new forms of knowledge is less unusual than we might like to think. It is not simply the novelty or legitimacy of ideas that may unsettle the ruling elites or class of an era. Paradigmatic changes in the way we collectively think about and understand the world have potentially transformative implications.[24] The clerical establishment that wielded such power in Copernicus' time denounced his ideas as heretical, although this owed more to a generalized repudiation of scientific ideas and was triggered by the Catholic Church's persecution of Galileo, an outspoken champion of heliocentrism.[25] The point of this short historical digression is that new ideas about the nature of reality have often proved threatening to those in power or who benefit from the existing order. In the Catholic Church's case, the likes of Copernicus and Galileo represented a fundamental challenge to their authority and that of the 'gospel truth': if the theological hierarchy could be wrong in their understanding of the most fundamental questions about the nature of material reality, they could be wrong about everything.

Given that religion continues to play a large role in individual belief systems, people's collective identities and the actions that theologically inclined states supposedly take on their behalf, this is plainly an issue that has proved surprisingly durable.[26] I say 'surprising' because many people, especially in the West, thought that religion would become less important as a source of personal belief, let alone collective action. Indeed, it was widely seen as a sign of collective 'progress' that inherited superstitions and belief systems would be discarded in favor of more rational and scientific explanations of reality. And yet many people not only continue to believe that a transcendental realm exists, but they also believe that this may have significance for quotidian reality.[27] Such belief systems clearly continue to inform their behavior and responses to many issues and may have more purchase than technical debates about the relative merits of alternative public policies.

Before considering just what those processes are, some final intellectual ground clearing is in order. First, many people's beliefs about and reactions to the very possibility of climate change also have decidedly nonscientific bases, but they are not exclusively religious. In part, this has come about as a consequence of the diminished status of scientific knowledge in particular and expertise more generally. An increasing number of people have little confidence in the ability of experts, scientific or otherwise, to provide truthful, objective accounts of climate change and much else.[28] As we shall see, the collapse in confidence of 'authoritative' sources of knowledge and the ability of elites to address problems—especially in non-self-serving ways—is one of the key factors that underpin the rise of populism, especially in the West. The dilemma is nicely summed up by Mark Brown, who argues that

> climate science rejection is not only an ideological tool for defending economic interests, a psychological defense mechanism, or a cowardly abandonment of reason and rationality. It may be all those things. But it is also part of a long tradition of popular distrust of power, in this case the power of science rather than government. And it is part of a tendency across the political spectrum to use science as a proxy battleground for politics. Rejectionism is not simply an unwillingness to face an 'inconvenient truth,' but a political reaction against those who would use truth to eliminate politics.[29]

The second point to make is subjective. At the outset of this book, I alerted the reader to the idea that it's impossible to have an 'objective' view about climate change, my enthusiasm for the merits of the scientific method notwithstanding. Even in the unlikely event that we can all agree on the underlying drivers and implications of climate change, our response to this issue is inevitably influenced by who, where and what we are. I am a child of the Western Enlightenment and still a believer in the possibility of Progress, especially the intellectual variety.[30] In this context, the *fact* that we now realize our planet is decidedly not the center of a clockwork universe in which the Earth occupies a unique place is an immensely important step forward in our understanding of our collective circumstances. This is progress on stilts as far as our species is concerned.

Does this have any wider philosophical, ontological significance or even theological significance? I'm afraid I have no idea. I'm an agnostic, which makes me something of a professional 'don't know'—not an encouraging confession for any reader hoping to discover how to save the world by reading this volume. One of the reasons for my agnosticism is that, despite the fact that 'we'—that is, the physicists who specialize in such things and their admirers—have good, widely accepted explanations of all that has followed the Big Bang (in empirical, cosmological terms, at least), we don't know what was going on before or why it happened. In other words, there are still major limits to our collective understanding. Life may have a meaning beyond the one we create for ourselves, but if it has, it's eluded me thus far, I'm afraid.[31] What is important for the current discussion, however, is that climate change is one of the phenomena we *do* know a lot about, and we can be pretty confident about the science, too—even if well-funded institutions have worked tirelessly to convince us otherwise.[32]

JUST THE FACTS, MA'AM[33]

The overwhelming majority of the scientific community working on climate change issues is in broad agreement about the nature of the drivers of global warming,[34] even if many lay people remain unaware of this. Despite the consensus about the nature of the problem and its causes, there is necessarily much less certainty about what this may mean in the future. At the outset, however, it is worth pausing to reflect that understanding something as complex as global warming is quite an achievement in itself, even if it is one that many people seem to be reluctant to take seriously.[35] There are plausible explanations for this apparent obduracy, and we need to be careful about simply dismissing climate change deniers as ignorant, ill-informed and/or shaped by world views that don't encourage critical thought, real dispassionate analysis or a willingness to take evidence seriously. More problematically, however, as we shall see, there are also some very smart people who seem to 'get' the evidence about climate change but choose to discount it or offer alternative responses that are not centered on trying to limit the increase of the 'greenhouse gases'—which *are* the central drivers of global warming and consequent climate change.[36]

Anyone who has been taking even a passing interest in the subject of global warming and climate change will need little reminding of some of the key features of changes in the environment that have sparked such alarm, and—it must be said—indifference, incredulity and opposition. The basic scientific story is well known though: greenhouses gases play a crucial role in absorbing long-wave radiation from the sun. The amount of greenhouse gases in the overall atmosphere is relatively small, and has varied over millennia, but at any particular time they play an outsized role in determining the temperature of the planet. Of the principal greenhouse gases in the atmosphere—which include water vapor, methane, nitrous oxide and ozone—carbon dioxide (CO_2) is the most important. Unfortunately, it has been building up in the atmosphere at an accelerating rate and this is a major part of our collective problems.[37]

One of the many paradoxes and complexities of the climate change story is that while the planet contains massive quantities of energy that is relatively cheap and accessible, actually using some forms is problematic. Oil and coal not only fueled the industrial revolution and transformed the way we live, work and even organize ourselves politically,[38] but they also released massive amounts of CO_2 (and other pollutants) into the atmosphere in the process. In addition to being valuable resources, oil, coal and trees are also crucial 'carbon sinks'. Actually taking advantage of the energy they contain releases more CO_2, and this has become a crucial part of the overall carbon cycle in which CO_2 is removed from the atmosphere through photosynthesis or by dissolving in the world's oceans and lakes. So great has the build-up of CO_2 in the atmosphere been, however, that the oceans are becoming more acidic as the CO_2 interacts with water molecules.[39]

These problems would be significant in isolation, but the planetary scale and interconnectedness of so many activities, especially those that affect the natural environment, has become increasingly apparent. Human beings are not just conscious of such processes, however: they are driving them. Indeed, it has become increasingly customary to describe the contemporary era as the 'Anthropocene', because our individual and collective actions are quite literally changing the planet.[40] James Lovelock's famous Gaia hypothesis—which suggests that the living and nonliving parts of the Earth form a complex interactive, single organism—captured this idea.[41] Whatever one may think of this notion, which has been subjected to some sustained criticism by evolutionary biologists like Richard Dawkins,[42] it does highlight the fact that the collective impact of human beings is having a major impact on the biosphere that sustains life for all living creatures on the planet.

There are, indeed, some interesting philosophical and even ethical questions about the possible obligations the dominant species may have toward the planet and its other life-forms.[43] Before considering any of those, though, it is necessary to spell out just what the implications of global warming and associated climate change may be. It's worth reemphasizing that predictions about the consequences of an unprecedented series of events and interactions that are anthropogenic rather than 'natural' (if we can actually separate the two, of course), is necessarily imprecise and judgments vary as a consequence. The bottom line, however, is that it doesn't look good if we carry on with business as usual, and the following discussion is consequently rather gloom-inducing.

The End of Civilization as We Know It?

One of the problems confronting those of us who would like to encourage action before it may become too late is that many policymakers are either unwilling or incapable of doing so, and many of our fellow citizens appear blithely unconcerned or disbelieving. Consequently—and rather problematically for the central argument of this book—the chances of a groundswell of popular opinion compelling politicians to act are probably not good anywhere, or are not in pursuit of 'progressive' public policies, at least. Authoritarian regimes—whether populist or not—are capable of suppressing dissent and ideas that they don't like or which threaten the status quo. Although there is no straightforward relationship between illiberal forms of politics and any particular environmental regime, as we'll see in the case of China, hopes about the transformative effects of social media look overblown.[44]

Even in long-standing democracies, such as Australia, political mobilization faces an equally daunting challenge: some people are sick of hearing about the growing problems associated with climate change and appear to be simply switching off.[45] Such an outcome is rather deflating, but entirely understandable: the idea that catastrophic climate change could occur has been around for a couple of decades, but for many living in places like Australia, the US, even China, the impacts of global warming have not been great thus far. Yet there is evidence that this may be beginning to change even in politically and ideologically conservative rural communities in Australia, as making a living as a farmer in the world's driest continent becomes an even more precarious proposition.[46] So far, however, even seemingly dramatic increases in average temperatures and changes in rainfall distribution have done little to shift the policy debate. On the contrary, in the US and Australia in particular, politicians are very much sticking to business as usual. But as the old joke has it, something that looks like it can't go on forever, probably won't.

It is important to remember just how novel, unprecedented and—in the long-term history of humanity—recent this transformation in our collective relationship with the environment actually is: 'In little over two generations—or a single lifetime—humanity (or until very recently a small fraction of it) has become a planetary-scale geological force.'[47] What climate scientists describe as the 'great acceleration' of socioeconomic processes that have interacted with the biosphere only really took off after World War II. True, many of the technological and industrial ingredients that have enabled (some of) us to transform our lifestyles and living standards had been in place for decades, but it was only in the 1950s that a combination of rapid growth in population and economic development really began to impact on the natural environment. Some of the most important indicators of major and potentially catastrophic change—increased greenhouse gas emissions, temperature increases, ocean acidification, tropical forest loss, terrestrial biosphere degradation—have all changed dramatically for the worse.[48]

Yet part of the problem as far as any public policy urgency is concerned is that some of the most dramatic possible effects of global warming, such as rising sea levels caused by melting ice caps, may take decades to have a major impact. If the world proves incapable of acting in the way most climate scientists believe is absolutely necessary, however, it is 'very likely' according to the IPCC's exhaustive survey of the evidence and possible trends that by the end of the twenty-first century sea level rise will occur in 95 percent of the world's oceans.[49] Given the millions of people that live in coastal communities and the fact that some of the world's more significant cities will be potentially vulnerable to inundation, business as usual will clearly not be an option. Cities in Mozambique, Madagascar, Mexico, Venezuela, India, Bangladesh, the Philippines and Vietnam are judged to be particularly at risk,[50] but look especially ill-equipped to cope with such challenges. Indeed, the prospect of an entirely predictable catastrophe does *not* mean that politicians will necessarily spring into action and put plans in place to retrieve the situation. What it does mean is that the proverbial window of opportunity in which effective intervention might ward off the worst effects of climate change is rapidly closing.[51]

Not only does the amount of CO_2 in the atmosphere continue to rise, but future increases in greenhouse gases are also inevitable as there is a lag between emission and their impact on the environment. In other words, even if we stopped using fossil fuels tomorrow, climate change and warming are likely to continue for some time. This not a justification for despair or inaction, of course, but it is a reminder of the urgency of the situation. As Sackett and Steffen point out:

> Ten years ago, the IPCC concluded that in order to keep the average global temperature rise between 2.0 and 2.4 degrees Celsius at the end of the century, global emissions would need to peak between the years 2000 and 2015, falling rapidly thereafter, and thereby holding atmospheric CO2 concentration between 350 and 400 parts per million (ppm). And yet, at the end of 2018, global greenhouse gas emissions continue to rise, and the targeted CO2 stabilization range has been exceeded, with average yearly concentrations now 405 ppm or higher…The world has probably already missed the 1.5 degrees Celsius target.[52]

Given that the latest IPCC report on the possible impact of global warming is predicated on a figure of 1.5 °C, this is a sobering thought. There is a surprisingly large projected difference in limiting overall warming to 1.5 °C rather than 2 °C, but even the latter is better than some of the truly apocalyptic predictions associated with even higher temperature increases.[53] As it is, even a best-case scenario rise of 'only' 1.5 °C will see significant declines in agricultural production, fish stocks and general health as a result of heat stress and increased exposure to various diseases, loss of ecosystems and biodiversity, water shortages and a general increase in poverty in some of the world's most vulnerable communities. When 'extreme weather events' of a sort that have become such a feature of recent times are added to the mix, the picture is not a happy one. And, it is important to remember, this is all predicated on the idea that it may be possible to limit further global temperature rises to less than 2 °C![54]

The reader will, I hope, forgive the exclamation mark, but for once it seems merited. The consequences of climate change really are both unprecedented, at least as far as human beings are concerned, and potentially terminal. Certainly, human beings have the capacity, consciously or unconsciously, to eliminate other species and are doing so with increasing speed. The key question is whether we also have the capacity to recognize and rectify our impact on the environment upon which other creatures besides ourselves depend.[55] As we shall see in subsequent chapters, there are some encouraging developments, even if the unprecedented scale and nature of the adjustments in combination with a forbidding timeframe and a lackluster historical record of effective policymaking do not inspire complete confidence.

In our collective defense, however, it is worth making a couple of points. First, it needs to be reemphasized that global warming presents a challenge of a completely unprecedented nature. The technical and political problems are consequently immense; it is no surprise, perhaps, that some of our leaders are floundering, in denial or simply flummoxed. Even at the best of times, it is not easy to encourage effective international cooperation, something the rise of authoritarian populism makes even less likely, as we shall see. The second point to make is that 'we' didn't set out to wreck the planet, inflict ecocide on our fellow creatures, or generally stuff things up. Economic development and even consumption are inherently desirable things; pursuing them seemed like a good idea. It still does. Any politician who suggests otherwise invites criticism at best, overthrow at worst. The legitimacy and authority of the People's Republic of China's (PRC) 'communist' leadership, for example, is largely dependent on its continuing ability to keep delivering rising living standards and economic stability.[56]

The question in this sort of context, therefore, is whether it is possible to have continuing economic expansion and rising living standards, especially when the overall global population is still growing. There are longstanding concerns about the carrying capacity of the planet, especially if the world's population copies the resource-intensive lifestyles of the developed world.[57] True, global demographics are changing, and world population growth is expected to slow, but it is still likely to reach 9.8 billion in 2050 and an astounding 11.2 billion by 2100.[58] The majority of this population growth is likely to occur in the so-called developing world, where the prospects for development look likely to remain remote, especially in more unforgiving environmental circumstances.[59] Indeed, one of the most difficult problems in achieving any sort of consensus on what might or should be done to combat climate change is that the world remains a very unequal place and some people are not only already more badly affected by global warming, but they understandably feel aggrieved about the possibility that they may never be able to replicate the developmental 'success' of the West. Put simply, given that the world's major decision-making processes are still dominated by the West, getting a grip on climate change may consign many of the world's poor—who are least responsible for the problem in the first place, of course—to continuing poverty and marginalization.[60]

Conclusion

Jared Diamond's brilliant analysis of civilizational decline as a consequence of human beings getting out of balance with the natural environment that ultimately supports them is a sobering reminder of all that can go wrong. Some of his conclusions are frankly alarming, especially for Australia: Diamond is 'doubtful' that Australia can support its present population, let alone the much larger numbers that some business people and strategic thinkers consider desirable if the country is to be a major force in international affairs. On a slightly cheerier note, Diamond thinks that humanity as a whole may be facing '"just" a future of significantly lower living standards, chronically higher risks, and the undermining of what we now consider some of our key values'.[61] Nothing to worry about there, then.

One of the important lessons that Diamond and others highlight is that environmental crises are nothing new, but part of the long-run history of human development.[62] While that may be true, what sets today's problems apart, of course, is the sheer scale of a challenge that has assumed planetary proportions with unambiguously global impacts. The social and economic developmental strategies that worked in an earlier era look simply unsustainable in the contemporary era. In the face of all-too-plausible predictions about threats to the existing order, however, there is little choice other than to attempt to respond, no matter how unlikely or implausible some of the responses may be. At a time of unprecedented change and potential crisis, even unlikely ideas and proposals may come to be taken more seriously.

One clear illustration of this is the growing interest in 'geoengineering'.[63] The significance of geoengineering is that many people hope that some technological solution will emerge that will not only 'save the planet', but will preserve many of the social political and economic structures that are currently in place, despite the fact that some of them seem to be directly implicated in our current problems. Indeed, Clive Hamilton argues that 'the promise of geo-engineering is the perfect excuse for decades of delay.'[64] Perhaps technology will save the day, but it would be unwise to bank on it; or to persist with business as usual in the expectation that it will, or that nothing really needs to change. Nothing better illustrates the perils and possibly unfulfillable promise than the economic system.

Notes

1. Olson, M. (2009) *The Logic of Collective Action: Public Goods and the Theory of Groups* (Cambridge, Mass.: Harvard University Press), p. 2. [Emphasis in original].
2. Ostrom, E. (1990) *Governing the Commons: The Evolution of Institutions for Collective Action* (Cambridge: Cambridge University Press), 6.
3. Whitmarsh, L. (2011) 'Scepticism and uncertainty about climate change: Dimensions, determinants and change over time', *Global Environmental Change* 21(2): 690–700.
4. Fukuyama, F. (2017) 'The emergence of a post-fact world'. *Project Syndicate* January 12.
5. The literature detailing the nature and possible impact of climate change is now extensive and generally considered credible. Some of the more important contributions include Intergovernmental Panel on Climate Change (2007) *Climate Change 2007: The Physical Science Basis* (Geneva: IPCC Secretariat); Stern, N., Peters, S., Bakhshi, V., Bowen, A., Cameron, C., Catovsky, S., Crane, D., Cruickshank, S., Dietz, S. and Edmonson, N. (2006) *Stern Review: The Economics of Climate Change*, vol. 30 (London: Cambridge University Press); Garnaut, R. (2011) *The Garnaut Review 2011: Australia in the Global Response to Climate Change* (Melbourne: Cambridge University Press).
6. Mueller, J.E. (1989) *Retreat From Doomsday: The Obsolescence of Major War* (New York: Basic Books); Tannenwald, N. (2007) *The Nuclear Taboo: The United States and the Non-use of Nuclear Weapons Since 1945* (Cambridge: Cambridge University Press).
7. United Nations Development Programme (2007) *Fighting Climate Change: Human Solidarity in a Divided World* (New York: UNDP).
8. Roberts, J.T. and Parks, B.C. (2007) *A Climate of Injustice: Global Inequality, North-South Politics, and Climate Policy* (Cambridge, Mass.: MIT Press).
9. The assumed need for independent military strength and capability is a fundamental component of strategic thinking in the major powers, despite some differences in strategic cultures. See Johnston, A.I. (1995) 'Thinking about strategic culture', *International Security* 19(4): 32–64.
10. Sokolski, H.D., ed. 2004. *Getting MAD: Nuclear Mutual Assured Destruction, Its Origins and Practice* (Darby: Diane Publishing).
11. Campbell, K.M., Gulledge, J., McNeill, J.R., Podesta, J., Ogden, P., Fuerth, L., Woolsey, R.J., Lennon, A.T., Smith, J. and Weitz, R. 2007. *The*

Age of Consequences: The Foreign Policy and National Security Implications of Global Climate Change (Washington: Center for Strategic and International Studies).
12. Winch, P. (2002) *The Idea of a Social Science: And Its Relation to Philosophy* (London: Routledge).
13. There is a large literature on the 'social construction of reality' and the way we think about the world. See Berger, P. and Luckmann, T. (1967) *The Social Construction of Reality: A Treatise in the Sociology Knowledge* (London: Penguin); Ruggie, J.G. (1998) 'What makes the world hang together? Neo-utilitarianism and the social constructivist challenge', *International Organization* 52(4): 855–885.
14. Oreskes, N. (2004) 'The scientific consensus on climate change', *Science* 306(5702): 1686–1686; Cook, J., Nuccitelli, D., Green, S. A., Richardson, M., Winkler, B., Painting, R., Skuce, A. (2016). 'Comment on "Quantifying the consensus on anthropogenic global warming in the scientific literature.' *Environmental Research Letters*, 11(4).
15. Bringel, B.M. and Domingues, J.M. (eds. 2015). *Global Modernity and Social Contestation* (London: Sage).
16. Barber, B.R. (2001) *Jihad vs. McWorld* (New York: Ballantine Books); Hall, P.A. and Soskice, D. (eds. 2001) *Varieties of Capitalism: The Institutional Foundations of Comparative Advantage* (Oxford: Oxford University Press).
17. Francis Fukuyama famously made the most sweeping claim about the possible course of history, but he was far from alone in thinking that globalization was likely to compel states to subscribe to similar sorts of public policies. See Fukuyama, F. (1992) *The End of History and the Last Man* (New York: Free Press); Friedman, T.L. (2006) *The World is Flat: A Brief History of the Twenty-first Century* (New York: Macmillan); Dobbin, F., Simmons, B. and Garrett, G. (2007) 'The global diffusion of public policies: Social construction, coercion, competition, or learning?', *Annual Review of Sociology* 33, 449–472.
18. Fukuyama, F. (2018) *Identity: Contemporary Identity Politics and the Struggle for Recognition* (London: Profile).
19. Murphy, C.N. (2000) 'Global governance: Poorly done and poorly understood', *International Affairs* 76(4): 789–803.
20. Atteridge, A., Shrivastava, M.K., Pahuja, N. and Upadhyay, H. (2012) 'Climate policy in India: what shapes international, national and state policy?', *Ambio* 41(1): 68–77.
21. Paty, M. (1999) 'Comparative history of modern science and the context of dependency', *Science, Technology and Society* 4(2): 171–204.
22. Hankins, T.L. (1985) *Science and the Enlightenment* (Cambridge: Cambridge University Press).

23. Although C. Wright Mills used this term to describe a small group of powerful people that dominated decision-making in the US, the idea captures something important about a broader historical pattern that has not disappeared. See, Mills, C.W. ([1956] 2000) *The Power Elite*, Inequality (Oxford: Oxford University Press); Rothkopf, D. (2008) *Superclass: The Global Power Elite and the World They Are Making* (New York: Farrar, Straus and Giroux).
24. Hall, P.A. (1993) 'Policy paradigms, social learning, and the state: The case of economic policymaking in Britain', *Comparative Politics* 25(3): 275–296.
25. Langford, J.J. (1992) *Galileo, Science, and the Church* (Ann Arbor: University of Michigan Press).
26. Herbert, D. (2017) *Religion and Civil Society: Rethinking Public Religion in the Contemporary World* (London: Routledge).
27. Emerson, M.O. and Hartman, D. (2006) 'The rise of religious fundamentalism', *Annual Review of Sociology* 32, 127–144.
28. Collins, H. and Evans, R. (2008) *Rethinking Expertise* (Chicago: University of Chicago Press).
29. Brown, M.B. (2014) 'Climate science, populism, and the democracy of rejection', in, D.A. Crow and M.T. Boykoff *Culture, Politics and Climate Change: How Information Shapes Our Common Future* (London: Routledge), p. 141.
30. Stephen Pinker gives one of the most powerful statements in support of continuing progress in human affairs, even if his discussion of the environment is comparatively thin and not entirely convincing. See, Pinker, S. (2018) *Enlightenment Now: The Case for Reason, Science, Humanism, and Progress* (New York: Viking).
31. It's important to note that some prominent atheists are especially scathing about organized religion and this makes the possibility of effective and inclusive dialogue more difficult. See, Dawkins, R. (2016) *The God Delusion* (New York: Random House).
32. One of the most notorious examples of this possibility is the George C. Marshall Institute, which despite professing to be interested in promoting scientifically informed 'sound' policymaking, has tried to cast doubt on the evidence for climate change. Oreskes and Conway have detailed how this Cold War creation turned its attention to climate change and the promotion of 'a radical free market ideology opposing any kind of restriction on the purist of market capitalism, no matter the justification.'
33. This is a rather dated, no doubt patronisingly sexist, reference to a phrase supposedly used by the fictional hero of the early television crime series *Dragnet*. The fact that it was apparently never uttered is a reminder of how historical 'facts' can take on a life of their own, and remind us of the social

circumstances that created them. Oreskes, N. and Conway, E.M. (2008) 'Challenging knowledge: How climate science became a victim of the cold war', in R.N. Proctor and L. Scheibinger *Agnatology: The Making and Unmaking of Science* (Stanford: Stanford University Press), p. 77.
34. Cook, J., Nuccitelli, D., Green, S.A., Richardson, M., Winkler, B., Painting, R., Way, R., Jacobs, P. and Skuce, A. (2013) 'Quantifying the consensus on anthropogenic global warming in the scientific literature', *Environmental research letters* 8(2): 1–7.
35. Nuccitelli, D. (2018) 'American conservatives are still clueless about the 97% expert climate consensus'. *The Guardian* 5 April.
36. Unfortunately, one of them is currently the most powerful man in the world. Dawsey, J. (2018) 'Trump on climate change: "People like myself, we have very high levels of intelligence but we're not necessarily such believers".' *Washington Post* November 27.
37. There are many excellent, scientifically informed explanations and analyses of these processes by now. See, for example, Garnaut (2011), *op cit*.
38. Timothy Mitchell argues that 'modern mass politics was made possible by the development of ways of living that used energy on a new scale.' See, Mitchell, T. (2011) *Carbon Democracy: Political Power in the Age of Oil* (London: Verso Books), p. 12.
39. Hoegh-Guldberg, O., Mumby, P.J., Hooten, A.J., Steneck, R.S., Greenfield, P., Gomez, E., Harvell, C.D., Sale, P.F., Edwards, A.J. and Caldeira, K. (2007) 'Coral reefs under rapid climate change and ocean acidification', *Science* 318(5857): 1737–1742.
40. Bonneuil, C. and Fressoz, J.-B. (2016) *The Shock of the Anthropocene: The Earth, History and Us* (London: Verso Books).
41. Lovelock, J. (2006) *The Revenge of Gaia: Why the Earth if Fighting Back – And How We Can Still Save Humanity* (London: Allen Lane).
42. Bond, M. (2013) 'Exploring our love/hate relationship with Gaia', *New Scientist*, August 23.
43. Singer, P. (2017) 'All animals are equal', in P. Clare *Animal Rights* (London: Routledge): 3–16.
44. Feldstein, S. (2019) 'How artificial intelligence is reshaping repression', *Journal of Democracy*, 30(1): 40–52; Feng, E. (2018) 'China and the world: how Beijing spreads the message'. *Financial Times* July 12.
45. Morrison, M., Parton, K. and Hine, D.W. (2018) 'Increasing belief but issue fatigue: Changes in Australian household climate change segments between 2011 and 2016', *PloS one* 13(6): 1–18.
46. Cox, L. (2018) 'Drought-stricken farmers challenge Coalition's climate change stance in TV ad'. *The Guardian* 16 Sep.
47. Steffen, W., Broadgate, W., Deutsch, L., Gaffney, O., & Ludwig, C. (2015). The trajectory of the Anthropocene: The great acceleration. *Anthropocene Review*, 2(1), p. 94.

48. Steffen et al., ibid., p. 87.
49. Intergovernmental Panel on Climate Change (2014) *Climate Change 2014: Synthesis Report* (New York: IPCC), p SYR 23.
50. Potsdam Institute (2012) *Turn Down the Heat: Why a 4 °C Warmer World Must Be Avoided* (New York: World Bank), p. 34.
51. Figueres, C. (2017) 'Three years to safeguard our climate', *Nature*, 546: 593–95.
52. Sackett, P.D. and Steffen, W. (2018) 'Don't feed the fossil fuel elephant'. *APPS Policy Forum* December 7.
53. Lynas, M. (2008) *Six Degrees: Our Future on a Hotter Planet* (London: Harper).
54. Intergovernmental Panel on Climate Change (2018) *Global Warming of 1.5 °C* (New York: IPCC).
55. Leakey, R.E. and Lewin, R. (1995) *The Sixth Extinction: Patterns of Life and the Future of Humankind* (New York: Doubleday).
56. Yang, H. and Zhao, D. (2015) 'Performance legitimacy, state autonomy and China's economic miracle', *Journal of Contemporary China* 24(91): 64–82.
57. Arrow, K., Bolin, B., Costanza, R., Dasgupta, P., Folke, C. and Mäler, K.-G. (1996) 'Economic growth, carrying capacity, and the environment', *Environment and Development Economics* 1(1): 104–110; Cohen, J.E. (1995) *How Many People Can the Earth Support* (New York: W.W. Norton); Rees, W.E. (2002) 'Globalization and sustainability: Conflict or convergence?', *Bulletin of Science, Technology & Society* 22(4): 249–268.
58. UD Department of Economic and Social Affairs (2017) *World population projected to reach 9.8 billion in 2050, and 11.2 billion in 2100*, 21 June, https://www.un.org/development/desa/en/news/population/world-population-prospects-2017.html
59. Solomon, N., Birhane, E., Gordon, C., Haile, M., Taheri, F., Azadi, H. and Scheffran, J. (2018) 'Environmental impacts and causes of conflict in the Horn of Africa: A review', *Earth-Science Reviews* 177, 284–290.
60. UNDP 2007, *Fighting Climate Change, op cit.*
61. Diamond, J. (2005) *Collapse: How Societies Choose to Fail or Survive* (London: Allen Lane), p. 7.
62. Ponting, C. (1991) *A Green History of the World: The Environment and the Collapse of Great Civilizations* (New York: Penguin).
63. Wigley, T.M.L. (2006) 'A combined mitigation/geoengineering approach to climate stabilization', *Science* 314(5798): 452–454.
64. Hamilton, C. (2010) *Requiem for a Species: Why We Resist the Truth About Climate Change* (London: Earthscan), p. 186.

CHAPTER 2

The Paradoxes of Development

If anything is to be done about placing the world on a more sustainable ecological pathway, changes in the way we collectively produce the things we consume will have to be part of the solution. At present, our overall impact on the finite resources of the planet is running down the 'natural capital' upon which we and—even more importantly perhaps—future generations depend. A key question to consider is whether the dominant economic model with which we organize production and consumption is compatible with long-term sustainability, or whether even the most 'civilized' and regulated forms of capitalism are actually the source of many of our problems. Are the kinds of consumption patterns associated with market economies and the pursuit of individual interests part of the problem rather than the solution?

There is a fundamental paradox at the heart of 'successful' economic development that will not be easily resolved. Although capitalism has proved uniquely useful in satisfying the material wants of at least some members of the human race, it has done so at enormous cost to the environment. Compounding this problem is the fact that the world's poor have not benefitted from the remarkable transformation in living standards and lifestyles that has been enjoyed by the comparatively fortunate few in the richer, 'North' of the global economy. As a result, any economic model that enjoys the support of the world's population will have to address the fundamental disparities of wealth and opportunity that have characterized the spread of global capitalism. As we shall see, however, the very models we use and the concepts with which we understand our collective relationship with the environment are not simply limited, but may actually be contributing to the problems we need to remedy.

It's Just Not Fair

The world is unequal, uneven and unfair. The last point is inevitably something of a value judgment, but there are plenty of people, and not just in the so-called developing economies or the more fashionable 'emerging markets', who would agree.[1] Global income and wealth distribution is becoming more skewed than ever and the chances of the 'bottom billion' escaping from grinding poverty, let alone replicating the privileged lifestyles of those of us living in the 'developed' world, are as remote as ever.[2] This is potentially a major handicap when trying to develop collective responses to transnational problems. If the current international order is judged to be fundamentally unjust and inequitable,[3] it is difficult to imagine the poor—or their notional representatives—signing up to a policy agenda that may have the effect of locking in the status quo.

Making sense of global development indicators is almost as complex as trying to understand the manifold impacts of global warming, but a few points are relatively uncontroversial. First, the poor—as God's son supposedly predicted—are still with us. While the number of people living in absolute poverty as a proportion of the world's population may have actually decreased from 11 to 10 percent, largely thanks to the 'East Asian miracle' and the economic rise of China, that still means that 736 million people live on less than $1.90 per day.[4] Significantly, many of the most disadvantaged are in sub-Saharan Africa, which has the misfortune of having both endemic poverty and some of the most rapid population increases in the world. The idea that a large population will deliver a demographic dividend is far from clear in Africa's case, especially when much of the continent is experiencing environmental stress.[5] More controversially, it is instructive to contrast the position of some African states with the Chinese experience, where the latter's much decried one-child policy has resulted in something like 400 million people not being born and consequently not contributing to China's (and the world's) environmental problems.[6] For all the current anxiety about China's ability to cope with an aging population,[7] it is also clear that population control was no impediment to economic development.

While China's experience may demonstrate that it is possible to overcome profound historical disadvantage and turmoil,[8] this will come as little comfort to those elsewhere hoping to experience the sorts of lifestyles currently enjoyed by China's rapidly expanding middle class. The point to emphasize is that capitalist development has been a highly 'uneven' process, and in some ways China's experience in particular and the East Asian region more generally is still exceptional in its speed and extent.[9] For the purposes of this discussion, however, what is most significant about China's ascent is that it has come at a fearful cost to its own domestic environment[10]; recently these problems have been exported to the rest of the world, primarily in the form of greater CO_2 emissions.[11]

For other countries, the historical legacy of uneven development has proved much more difficult to overcome. In this context, the fact that some parts of the world have benefitted from capitalist development, while others have been 'exploited' and their chances for development constrained, is seen as an inescapable part of a flawed economic system. Neil Smith, for example, argues that uneven development needs to be seen 'not simply as a "gap" between more developed and less developed regions or a universal phenomenon, but as the systematic product of previous capitalist development and the fundamental promise of the future of capitalism'.[12] If this analysis is correct, then it not only means that economic development and nondevelopment may have path-dependent causes and consequences,[13] but that there may also be a causal connection between the two. In other words, it is built into the nature of capitalist development that there are persistent, structurally embedded patterns of advantage and disadvantage.[14]

One novel way of thinking about this relationship, which presents a potentially socially embedded obstacle to economic justice and cooperation, is the idea of 'unequal ecological exchange'. In a manner similar to so-called dependency theory's account of development failure in Latin America and Africa,[15] unequal ecological exchange suggests that there is an enduring structural quality to the relationship between the developed and the underdeveloped worlds, or between the more fashionable 'North' and 'South'.[16] But however we describe the wealthier and poorer parts of the planet, the idea that the former may benefit from, and contribute to, not only the latter's lack of economic development, but also to the growing environmental problems of the South, is an important one that is generally not taken seriously by mainstream economics. Yet there is plentiful evidence about the unequal and exploitative relationship that has existed between developed and developing states.[17]

Japan, for example, casts a long environmental shadow over much of the Pacific, where its multinational corporations have systematically exploited the resources of its poorer neighbors, leaving Japan's own natural environment in much better shape as a consequence.[18] Similarly, while China's domestic environment may not have benefitted in quite the same way as Japan's, there is little doubt that its ecological footprint and impact is having an increasingly significant impact on the region of which it has become the most important economic actor. Significantly, it threatens to not only add to the pressures on the natural environment, but to lock poorer neighbors into unsustainable patterns of resource exploitation.[19]

While some may say it was ever thus and endemic to the nature of capitalist development, it raises important questions about the way we think about the utilization and exploitation of the natural environment that have generally not been given the attention they warrant in mainstream economic discourse—or the development policies of the influential international financial institutions (IFIs) that exert such an influence over the would-be developing world. The role of the IFIs is taken up in more detail in Chap. 3; the point to consider at this stage is whether a form of expansionary, resource-intensive capitalist development is actually any longer feasible given the finite nature of the planet.

CAN WE COPE WITH CAPITALISM?

Capitalism is now an economic system of truly planetary scope and scale. While there may important differences in the way it is actually organized and regulated in various parts of the world,[20] there is little doubt that all nations subscribe to economic principles that can be broadly described as capitalist. Private ownership, commodity production, consumerism and growth are central components of economic systems the world over. So far, so uncontroversial, perhaps. The question is whether capitalism is actually compatible with sustainable outcomes, be they narrowly economic or more broadly conceived issues of social justice. This question is especially true of the sort of 'neoliberal' capitalism that was promoted by Ronald Reagan and Margaret Thatcher during the 1980s. Neoliberalism is commonly seen as an especially pernicious Anglo-American creation that is associated with the winding back of state power and the concomitant 'intervention' in the markets of a sort that was associated with the managed economic development of the 'golden era' of capitalism.[21] Neoliberal theory, by contrast, is seen by critics as providing 'a benevolent mask full of wonderful-sounding words like freedom, liberty, choice, and rights, to hide the grim realities of the restoration of or reconstitution of naked class power, locally as well as transnationally, but most particularly in the main financial centres of global capitalism.'[22]

Even if some forms of capitalism are or were more 'civilized' than others, can they exist in conjunction with other, more brutal varieties, or be scaled up to provide a general template for sustainable economic organization? Simply positing such questions remains relatively novel but is arguably unavoidable if there is any possibility that large numbers of human beings can live in a sustainable relationship with the natural environment. There is, of course, a fundamental 'contradiction', to borrow a bit of Marxist jargon, at the heart of capitalist systems of production: while capitalism has an unparalleled capacity to generate wealth and consumption goods, it has generally done so at the expense of the natural environment and by depleting finite resources.[23] It has frequently been observed that we collectively need nearly two planets to offset our yearly consumption of nonrenewable resources,[24] even when only a minority of the world's population enjoys the sort of lifestyle that necessitates such intensive forms of resource exploitation. One widely held expectation is that as living (and by implication education) standards rise, individuals will come to demand better environmental standards, bringing about the best of all possible worlds.[25] While the evidence for this proposition is 'inconclusive',[26] even if it were true, as we have already seen, there are serious doubts about whether living standards in some of the poorest parts of the world can or will actually rise—especially in the time frames that are likely to influence environmental policymaking.

Unsurprisingly, perhaps, Marxist scholars in particular are deeply skeptical about the potential for capitalism to provide a sustainable basis for human society. For some observers, such as Joel Kovel, it is 'either capital or our future. If we value the latter, capitalism must be brought down and replaced with an ecologically worthy society'.[27] It is, however, worth noting that the environmental record of supposedly socialist societies is not encouraging either: China under the leadership of Mao Zedong took an entirely instrumental approach to the natural environment which established the preconditions for many of today's problems.[28] The Chernobyl disaster was emblematic of the Soviet Union's similarly cavalier attitude to environmental questions and management.[29] Whether or not any country has actually achieved something akin to socialism is a moot point, but it is fair to say that enhanced living standards and consumption patterns have been universal desiderata of whatever economic system has been adopted in most human societies thus far. It is capitalism's distinguishing feature that it is better at achieving this than any alternative yet seen; this is also why we find ourselves in such a precarious position environmentally, and as far as questions of economic justice are concerned. The result, as Peter Dauvergne points out, is that

> [n]ot only is environmentalism failing to produce sustainable patterns of global consumption, much of what policymakers in high-consuming economies are labeling as "environmental progress" is in reality little more than the wealthy world deflecting the consequences and risks into ecosystems and onto people with less power – and thus less influence over global affairs.[30]

It has been suggested that one of the reasons for the continuing exploitation of nonrenewable resources is not simply because many of the processes are often invisible, transnational and at arm's length, but because we collectively place too little value on the things we consume. One way of attempting to address this from a conceptual perspective has been by introducing the idea of 'ecosystems services', or the 'flows of materials, energy, and information from natural capital stocks which combine with manufactured and human services to produce human welfare'.[31] Put simply, the idea of ecosystem services is an attempt to put a value on all of the 'free' benefits we derive from the natural environment, such as water for agriculture, soil formation, nutrient recycling, pollination, biological control, raw materials, food production, genetic resources and—in 'normal' circumstances, at least—regulation of the composition of atmospheric gases. By trying to put an actual monetary value on the total contribution provided by the environment, the argument went, policymakers and business leaders in particular might come to recognize its worth and consequently be more inclined to protect it.

Well intentioned as this strategy may have been, it has been criticized for what was meant to be its key advantage: in the rush to appear 'pragmatic' and focused on hard-nosed economic reality, many ecologists and conservationists have replaced traditional valuations that privileged the intrinsic worth of nature itself, with one that subscribes to the judgment of the market.[32] Indeed, some critics have highlighted how impoverished the calculation of worth provided by mainstream economics actually is when it comes to making sense of and determining the worth of the natural environment. Clive Spash, for example, argues that

> the very irrelevance of modern economics as a means for understanding the functioning of the economy is why it can exist. It is harmless for key power interests, namely a political elite, the rich and multinational corporations. They can use its models and concepts as rhetorical devices when convenient and ignore them just as easily. Yet, the paucity of economic analysis has real implications because it supports claims such as: all is well with the world, there is such a thing as an efficient competitive economy, the rich deserve their wealth, corporations are a valid and good institution, markets supply freedom and economic growth will eradicate poverty.[33]

To be fair, there is now evidence to suggest that even in Australia and the US, where national governments are currently unenthusiastic or incapable of developing effective climate mitigation policies, at least some elements of broadly conceived 'business community' are actually beginning to realize that it is in their interests to lead the way[34]—even if this is partly the result of shareholder activism.[35] It is also important to recognize that, just as there is arguably no unambiguous, widely supported 'national interest', there is no single 'business interest' either. True, business in general looks to the state to ensure the sanctity of private property and the right to appropriate and dispose of any economic surplus generated from its activities as it sees fit, but it is also clear that there are major differences in attitudes toward public policy among economic elites. For example, the coal industry frets about 'stranded assets' and disinvestment if the environment movement becomes more powerful and effective, while entrepreneurs in the renewable sector see the potential to make money from an increasingly important industry.[36]

One of the reasons that phenomena such as shareholder activism and political pressure from 'below' are so potentially important is not simply because the status quo position of business as usual is unsustainable, but because some business interests are assiduously working to thwart needed change. The coal industry is, perhaps, the most notorious and high-profile offender in this context. Not only is the coal lobby well-funded and influential, but it has made itself part of the policymaking process. This strategy has been taken furthest in the US where the Trump administration has appointed former coal industry lobbyists to head the Environmental Protection Agency (EPA). Even more ominously, perhaps, the present incumbent, Andrew Wheeler, is also a climate change skeptic who has cast doubt on the credibility and independence of the IPCC.[37] The implications of such influence are not confined to domestic politics: at the 2018 international climate talks in Katowice, the Trump administration 'offered an unapologetic defense of fossil fuels.'[38]

The implications of business influence over the domestic and foreign policies of the US and other states will be taken up in subsequent chapters. The point to emphasize at this juncture is about the overall 'structural' power of business when it comes to setting national policy agendas.[39] Not only does the private sector in liberal democracies such as the US and Australia have the power to decide what kind of investment—and by implication economic activity—occurs, but it can also directly influence the policy discourse that determines the regulatory framework within which business operates. In the US in particular, lobbying politicians has become a lavishly funded and symbiotic part of the way business and politics are conducted, which helps to explain the growing economic inequality in that country.[40] Predictably enough, perhaps, some industries, such as tobacco and coal, have used sophisticated disinformation campaigns to cast doubt on scientific evidence and understanding of key public policy issues that could adversely affect their profitability.[41] Even more problematically in the long term, however, the political class in the US is increasingly compromised and incapable of acting independently. As Hacker and Pierson point out, 'members of Congress let lobbyists write bills not just because they get rewarded for doing so but because they have so little in-house expertise of their own'.[42]

A Planned Alternative?

The US is, perhaps, an increasingly extreme example of many of the pathologies that currently afflict democracies around the world.[43] There has recently been a plethora of books discussing the crisis of democracy in the US and across the world more generally.[44] The rise of populism is, of course, central to such concerns. But before considering whether any form of populism offers an alternative to the political and economic models that have been in place—in the West, at least—for half a century or so, it is important to say something about the different forms of capitalism that currently exist and ask whether any of them are capable of providing the basis for an environmentally sustainable economic order, or whether observers such as Spash, Kovel, Naomi Klein and others are correct in suggesting that *any* form of capitalism is part of the problem rather than the solution.

At first blush, the answer would seem to be unambiguously obvious: an expansionary system that is driven by individual rather than collective interests, and which derives its remarkable dynamism from increased production and consumption, looks inherently unsustainable. Marxist scholars in particular emphasize that 'though population growth and inappropriate technologies have played important roles in accelerating environmental degradation, the ecological rift we are now facing has its principal source in the economy'. The implication, Foster et al. argue, is that 'it is not a question of individuals seeking simply to withdraw from the capitalist economy but rather of creating a new ecological hegemony within civil society aimed at transforming the entire structure of production and consumption.'[45]

It is not hard to see why some feel that the 'imperial mode of living' enjoyed by a privileged minority primarily in the North is unsustainable and dependent on the continuing exploitation of both the environment and those in the South.[46] Even if this analysis is correct, however, to paraphrase Margaret Thatcher, is there any alternative? As we shall see, there are a number of populist leaders around the world who claim to have answers to many of the current problems that plague the world, but environmental issues are generally conspicuous by their absence or dismissed as fake news.[47] One country that *is* taking climate change seriously and trying to position itself as a leader in mitigation efforts,[48] not least because it sees commercial advantages in being ahead of the technological developmental curve, is China.[49] Equally importantly, China's leaders recognize that climate change generally and pollution in particular have the potential to generate dangerous social instability.[50]

The question to ask in this context is whether the Chinese model—or any other, for that matter—offers an alternative to the increasingly tarnished model of free market capitalism found in the US and Australia.[51] Even though many Chinese officials and academics feel uncomfortable about describing their economy as capitalist, there is little doubt that it is.[52] It is far from clear whether this will lead to any form of inevitable political transformation, much less a democratic revolution, however, as many observers suppose.[53] China's rising capitalist class seems more than happy to work closely with the Chinese Communist Party (CCP) as long as this does not interfere with its ability to make money.[54] On the contrary, there is a close nexus between economic and political elites in China that Xi Jinping's recent crackdown on corruption has done little to overturn.[55]

For the purposes of this discussion, the key question is whether China's distinctive model of 'state capitalism', which retains many of the features of its former centrally planned economy—especially regarding the dominant role of the CCP[56]—actually offers an alternative to the sort of neoliberal model that has attracted such criticism of late.[57] Given that one of the principal criticisms of the American model of capitalism and state-business relations is that it has been captured by vested interests and is consequently incapable of meaningful reform,[58] it is striking that both systems appear inflexible and at the mercy of unrepresentative elites. The pernicious, self-serving influence of the coal lobby in particular and 'Wall Street' more generally in the US has had a baleful impact on environmental policy, Donald Trump's promises to 'drain the swamp' notwithstanding.[59]

A striking contrast with China in this context is that, despite their increasingly authoritarian approach to domestic and foreign policy, the elites that dominate the policymaking process in the PRC have a capacity and willingness to make things happen in a way that America's gridlocked political system does not. The PRC leadership recognizes the potential dangers to social stability posed by environmental degradation and has made real efforts to do something about it. It is no coincidence that China is the biggest investor in renewable energy in the world or that it has compelled some polluting industries to either shut down or move to less politically sensitive areas.[60] It is also clear that there are limits to how fast and how far this process can be taken: economic growth remains the overwhelming policy priority, without which the CCP's dominant role and authority would inevitably be called into question. This is also why, despite the massive investment in renewables and nuclear energy, China's use of coal continues to increase along with its CO_2 emissions.[61]

Significantly, it is not only China's ecological footprint that is having an impact across national borders. China's rise and the nature of its current foreign policies reflect a complex set of interlocking, historically unique forces that shape its goals and even the nation's sense of itself.[62] It is always dangerous to generalize about a nation of more than 1 billion people, but as we shall see in subsequent chapters, national identity and the role of individual states remain important obstacles to international cooperation. Just as the US did in the aftermath of World War II, China is seeking to use its growing material importance and strategic leverage to pursue increasingly ambitious foreign policy goals.[63] The implications for possible cooperative governance are explored in Chap. 3; at this point two key points about China's increasingly influential developmental model are worth emphasizing.

First, the foreign, domestic and environmental policies of *any* nation are strongly shaped by the national social and political context from which they emerge. China is certainly no exception to this, which reminds us that for all its noteworthy differences, the PRC's political economy has some familiar dynamics, albeit with Chinese characteristics. The scale of China's ambitions and their potential impact on the region and world more generally are, however—if they can be realized successfully— unprecedented in their scope and ambition. The Belt and Road Initiative (BRI), which promises to make China the center of an intercontinental trade and production network is the quintessential expression of these ambitions.[64]

The second point that flows from this is not simply that China is actively exporting a developmental model that is energy-intensive and predicated on continuing economic expansion, but that it is locking some of its neighbors into dependent relations that increasingly revolve around the Chinese economy and its associated flows of trade and investment.[65] More troublingly, as Rao and Baruah point out, 'unfortunately, this has meant that some BRI partners have become markets for the relocation of China's low-valued processing units, becoming "environmentally unsustainable dumping grounds"'.[66] China has also demonstrated the limits of its so-called good neighbor policy as far as its own national priorities are concerned[67]: acute water shortages in parts of China have led it to adopt hard-nosed policies toward its immediate neighbors with which it shares key natural assets like the Mekong River. The geopolitical, developmental and ecological consequences are potentially dire:

> China is supremely confident of its position, with two countries – Laos and Cambodia – enmeshed in a nexus of loans, investment, and obligations already on board. But a simmering conflict over the equitable sharing of water resources is deeply felt in Thailand, and even more in the Vietnamese delta, where upstream dams and climate change have made the region more prone to severe drought… The threat posed by over-damming the Mekong, coupled with the impacts of climate change, should be on the radar of regional and international organizations. The consequences for Cambodia and Vietnam would be devastating and reverse much of the progress made toward meeting UN Sustainable Development goals.[68]

The reason for focusing so intently on China's developmental prospects and impact is threefold. First, China is a rising power that is widely expected to replace the US as the largest economy in the world; it has already replaced the US as the largest emitter of CO_2. It is, as we have seen, also having a direct impact on many of its neighbors and is contributing to both their environmental problems and their ability to do anything about them. Second, and despite its own formidable environmental challenges, for some observers, not least in China itself, the PRC offers an alternative political and economic model to the increasingly discredited neoliberal variety that is associated with many of the world's contemporary environmental problems.[69] Third, while not strictly a populist regime, perhaps, political activity in the PRC continues to revolve around the CCP and the increasingly dominant figure of Xi Jinping. Any truly global response to the challenge of global warming will have to incorporate China to have any chance of success, which is why the rest of the world has to reckon with China's authoritarian alternative.[70]

Are Other Pathways Possible?

Before considering whether China can actually be accommodated, much less integrated into any global environmental compact, it is important to ask whether the PRC leadership's underlying commitment to economic growth and development makes effective cooperation impossible. Unfortunately, there are good reasons to think that it does: even a more state-centric version of business as usual is likely to put unsustainable strains on the natural environment, not least in China itself, which is already struggling to cope. This is why some observers argue that the only long-term solution to the environmental crisis—if, indeed, there actually is one—is to embark on a radical process of 'degrowth', which is predicated on the assumption that endless economic expansion, consumption and utilization of resources is impossible on a finite planet. While this assumption looks entirely reasonable at one level, its proponents often make their claims in self-referential isolation from the much-invoked 'real word'.[71] However laudable the idea may be in theory, it is effectively a nonstarter in practice, or not in the political and economic conditions that currently prevail in the world, at least. As Hausknost points out,

The idea that all regional powers worldwide would voluntarily restrict themselves to the development and application of only convivial technologies is rather implausible. Instead, the more reasonable assumption is that technologies of domination, warfare and mass production will be available and employed in the future, at least by some. This alone creates power imbalances that need to be addressed by any serious conceptualisation of degrowth democracy.[72]

Short of a revolutionary transformation of the existing political, economic and social order—which cannot be ruled out given the scale and growing impact of environmental problems, of course—any policy response will have to work within the existing institutional and regulatory frameworks that currently shape the international system. Neither the leaders of China, America or anywhere else, for that matter, seem to be contemplating abandoning an economic order broadly based on the operation of market forces and the private appropriation of profit. As Newell points out, in such circumstances regulation is likely to be more effective and enforceable when it is seen as 'regulation *for* rather than *of* business.'[73] He may well be right, but such a conclusion means that the fate of humanity will largely be in the hands of an economic elite whose interests are unlikely to coincide with the poor, the powerless and the marginalized, even in their own countries, let alone the citizens of countries that occupy a subordinate or redundant place in the global economy.

It is striking that some of the most attractive-sounding and plausible visions of economic change and reform are predicated on the existing order, rather than a complete social transformation. Ideas about a 'Green New Deal', for example, which have recently gained political traction in the US,[74] may point the way forward to a low-carbon economy and job creation through the mass installation of solar panels, retrofitting coastal infrastructure, manufacturing electric vehicles and other environmentally friendly initiatives. One of the most detailed and persuasive explications of what a sustainable Green economy might look like, has been developed by Edward Barbier, and published in conjunction with the United Nations Environment Programme (UNEP).[75] In addition to spelling out the potential benefits of initiatives such as retrofitting buildings to improve energy efficiency, developing mass transit systems, 'smart' electricity grids, wind and solar power, and advanced biofuels, Barbier grapples with the complexities of funding sustainable and productive economic activity in those parts of the world that have not benefitted from the era of postwar development.

Such ideas have been around for some time, of course,[76] but they do potentially offer a path to a more sustainable environmental order, including in the 'advanced' economies, which is why there is growing interest in such ideas in the US. The question will be whether such initiatives could be scaled up to include less 'developed' parts of the world, or whether exclusively national priorities would continue to shape the policy agendas of the most powerful states, and the economic elites who currently exercise such a disproportionate influence over our collective fate. While the likes of Barbier are optimistic about the role that organizations such as the G20 can play in promoting new ideas and modes of production, as we shall see in Chap. 3, there are still some formidable obstacles to cooperation, not least in the G20 itself.[77] As the Australian case reminds us, even at the national level, there are formidable political and economic obstacles to actually implementing what are widely regarded as necessary policies,[78] such as a tax on carbon.[79] Nevertheless, the fact that credible alternatives to the prevailing business-as-usual model actually exist at all is encouraging.

It is, however, inevitable that in a capitalist economy the controllers of capital exercise an outsized influence on the way we both organize and *think* about economic activity. The sanctity of private property—and thus control of the actual material environment itself—is a social construction that is enshrined in elaborate legal systems and ultimately dependent on individual states to protect and enforce. It is unsurprising that of the four main environmental perspectives described by Clapp and Dauvergne 'market liberalism' is the most familiar and influential. The challenge for alternative visions, especially those Clapp and Dauvergne call 'bioenvironmentalists' and 'social greens', is that their proposed solutions represent radical transformations in the way societies and economies are currently organized. No matter how accurate their analysis of the problems may be and how appropriate their suggested remedies, it is difficult to imagine them being adopted in the absence of a major 'critical juncture'[80] or—more feasibly, perhaps—irresistible pressure from below. Both the issues are explored in more detail in subsequent chapters.

Conclusion

This brief review of the paradoxes of development highlights a number of issues that are central to the difficulties we collectively confront. First, lifting people out of poverty is no small achievement and is difficult to argue against. The problem, of course, is that we have yet to discover a better way of doing this than by using some form of growth-oriented, market-driven, expansionary model of development. One possible reading of China's transition from centrally planned 'socialism' under Mao Zedong to the economic liberalism ushered in by Deng Xiaoping is that individual self-interest is clearly a more powerful spur to economic development than any sense of collectivism. Dispiriting as this claim may be to those on the 'Left'—if such labels actually mean anything any longer—it is yet another potential obstacle to resolving collective action problems that will have to be taken seriously.

The second obvious and potentially insurmountable problem that emerges from an analysis of the contemporary economic order is that it is inflicting an enormous toll on the natural environment. It is possible that *any* form of economic order that privileged the satisfying of material wants would inevitably run into the same sorts of problems, of course, as the experiences of China and the Soviet empire remind us. The challenge is to marry personal values and identities with a heightened consciousness of what our individual actions add up to as far as the environment is concerned.[81] It is for this reason that many 'deep ecologists' argue that what is needed is nothing less than a transformation in the way human beings think about themselves and the natural environment in a 'project of profound cultural remaking and renewal' that overcomes the 'illusory sense of our independence from nature.'[82] Perhaps so, but in the timeframes available to us to actually address immediate problems, it looks as if some version of the extant political order will have to do the heavy-lifting. Thus far, the international dimension of this order doesn't inspire great confidence.

Notes

1. Milanovic, B. (2016) *Global Inequality: A New Approach for the Age of Globalization* (Boston: Harvard University Press).
2. Collier, P. (2007) *The Bottom Billion: Why the Poorest Countries Are Failing and What Can Be Done About It* (Oxford: Oxford University Press).
3. Beitz, C. (2001) 'Does global inequality matter?', in, T.W. Pogge *Global Justice*. (Molden, Mass: Blackwell): 106–122.
4. World Bank (2018) 'Decline of Global Extreme Poverty Continues but Has Slowed', *press release*, September 19. https://www.worldbank.org/en/news/press-release/2018/09/19/decline-of-global-extreme-poverty-continues-but-has-slowed-world-bank
5. Pilling, D. (2018) 'Africa cannot count on a demographic dividend'. *Financial Times*, August 16.

6. Parkinson, J. (2015) 'Five numbers that sum up China's one-child policy, *BBC News Magazine*, 29 October, https://www.bbc.com/news/magazine-34666440
7. Chan, M. and Zhuang, P. (2018) 'China's ageing population problem worsens as birth and marriage rates fall'. *South China Morning Post* 16 July.
8. China suffered imperial predation, dynastic collapse, civil war and foreign invasion during its so-called '100 years of shame', and yet eventually experienced the most rapid large scale development process the world has ever seen. See, Callahan, W.A. (2010) *China: The Pessoptimist Nation* (Oxford: Oxford University Press).
9. Beeson, M. (2014) *Regionalism and Globalization in East Asia: Politics, Security and Economic Development*, 2nd ed. (Basingstoke: Palgrave).
10. Economy, E.C. (2004) *The River Runs Black: The Environmental Challenge to China's Future* (Ithaca: Cornell University Press).
11. Hornby, L. (2018) 'China's carbon emissions set for fastest growth in 7 years'. *Financial Times*, May 30.
12. Smith, N. (2008) *Uneven Development: Nature, Capital, and the Production of Space*, Third ed. (London: Verso), p. 207.
13. Path dependency refers to the possibility that certain patterns of behaviour and development may be locked in or rendered much more likely by their embedded, institutionalised nature. See, Campbell, J.L. (2004) *Institutional Change and Globalization* (Princeton: Princeton University Press).
14. Pierson, P. (2000) 'Increasing returns, path dependence, and the study of politics', *American Political Science Review* 94(2): 251–267.
15. Heller, P., Rueschemeyer, D. and Snyder, R. (2009) 'Dependency and development in a globalized world: Looking back and forward', *Studies in Comparative International Development* 44(4): 287–295.
16. Hornborg, A. (2001) *The Power of the Machine: Global Inequalities of Economy, Technology, and the Environment* (Walnut Creek: Rowman Altamira).
17. The literature is vast, but see Reinert, E.S. (2007) *How Rich Countries Got Rich…And Why Poor Countries Stay Poor* (New York: Carrol & Graf); Anievas, A. and Nisancioglu, K. (2015) *How the West Came to Rule: The Geopolitical Origins of Capitalism* (London: Pluto Press).
18. Dauvergne, P. (1997) *Shadows in the Forest: Japan and the Politics of Timber in Southeast Asia* (Cambridge, Mass.: MIT Press).
19. Coxhead, I. (2007) 'A new resource curse? Impacts of China's boom on comparative advantage and resource dependence in Southeast Asia', *World Development* 35(7): 1099–1119.
20. Hall, P.A. and Soskice, D. (eds. 2001) *Varieties of Capitalism: The Institutional Foundations of Comparative Advantage*. (Oxford: Oxford University Press).

21. Glyn, A., Hughes, A., Lipietz, A. and Singh, A. (1990) 'The Rise and fall of the Golden Age', in, S. Marglin and J. Schor *The Golden Age of Capitalism: Reinterpreting the Postwar Experience* (Clarendon Press): 39–125.
22. Harvey, D. (2007) *A Brief History of Neoliberalism* (Oxford: Oxford University Press), p. 119.
23. Klein, N. (2014) *This Changes Everything: Capitalism vs. the Climate* (London: Penguin).
24. Nace, T. (2017) 'Humanity has officially consumed more than Earth can produce this year', *Forbes*, August 3, https://www.forbes.com/sites/trevornace/#e50754531806
25. Stern, D.I. (2004) 'The rise and fall of the environmental Kuznets curve', *World Development* 32(8): 1419–1439.
26. Dinda, S. (2004) 'Environmental Kuznets Curve hypothesis: A survey', *Ecological Economics* 49(4): 431–455.
27. Kovel, J. (2007) *The Enemy of Nature: The End of Capitalism or the End of the World?* 2nd ed. (London: Zed Books), p. 159.
28. Shapiro, J. (2001) *Mao's War Against Nature: Politics and the Environment in Revolutionary China* (Cambridge: Cambridge University Press).
29. Ziegler, C.E. (1990) *Environmental Policy in the USSR* (Amherst: University of Massachusetts Press).
30. Dauvergne, P. (2010) *The Shadows of Consumption: Consequences for the Global Environment* (Cambridge, Mass.: MIT Press), p. 215.
31. Costanza, R., d'Arge, R., De Groot, R., Farber, S., Grasso, M., Hannon, B., Limburg, K., Naeem, S., O'neill, R.V. and Paruelo, J. (1997) 'The value of the world's ecosystem services and natural capital', *Nature* 387(6630), p. 254.
32. Gómez-Baggethun, E. (2017). Ecosystem services. In C. L. Spash (Ed.), *Routledge Handbook of Ecological Economics: Nature and Society* (London: Routledge): 445–454.
33. Spash, C.L. (2017) 'Social ecological economics', in, C.L. Spash *Routledge Handbook of Ecological Economics*. (London: Routledge), p. 6.
34. Coorey, P. and Ludlow, M. (2018) 'Business to go it alone on energy, climate policy'. *Australian Financial Review*, 10 Oct.
35. Rauf, D.S. (2018) 'Powerful investors push big companies to plan for climate change'. *Scientific American* May 3.
36. The regulatory setting and incentives provided by individual governments are crucial in this context. Yeates, C. and Latimer, C. (2018) "Greater uncertainty': Renewable investment boom tipped to slow'. *Sydney Morning Herald* 11 October.
37. Davenport, C. (2018) 'How Andrew Wheeler, the new acting E.P.A. chief, differs from Scott Pruitt'. *New York Times* July 5.

38. Plummer, B. and Friedman, L. (2018) 'Trump team pushes fossil fuels at climate talks'. *New York Times* December 10.
39. It is not only Marxists who recognise the power of business relative to other actors in society, so, too, do pluralists who worry about the 'capture' of the political class by powerful vested interests. See, Lindblom, C.E. (1977) *Politics and Markets: The World's Political-Economic Systems* (New York: Basic Books); Fuchs, D.A. (2007) *Business Power in Global Governance* (Boulder: Lynne Rienner Boulder); Bell, S. (2012) 'The power of ideas: The ideational shaping of the structural power of business', *International Studies Quarterly* 56(4): 661–673.
40. Hacker, J.S. and Pierson, P. (2010) *Winner-Take-All Politics: How Washington Made the Rich Richer – and Turned Its back on the Middle Class* (New York: Simon & Schuster).
41. Dunlap, R.E. and McRight, A.M. (2011) 'Organized climate change denial', in, J.S. Dryzek, R.B. Norgaard and D. Schlosberg *The Oxford Handbook of Climate Change and Society.* (Oxford: Oxford University Press): 144–160; Oreskes, N. and Conway, E.M. (2010) *Merchants of Doubt: How a Handful of Scientists Obscured the Truth on Issues from Tobacco Smoke to Global Warming* (New York: Bloomsbury Press).
42. Hacker, J.S. and Pierson, P. (2017) *American Amnesia: How the War on Government Led Us to Forget What Made America Prosper* (New York: Simon and Schuster), p. 344.
43. Levitsky, S. and Ziblatt, D. (2018) *How Democracies Die: What History Reveals About Our Future* (London: Viking).
44. Two of the best are Runciman, D. (2018) *How Democracy Ends* (London: Profile Books); Kuttner, R. (2018) *Can Democracy Survive Capitalism?* (New York: W.W. Norton).
45. Foster, J.B., Clark, B. and York, R. (2010) *The Ecological Rift: Capitalism's War on the Earth* (New York: Monthly Review Press), pp. 377 & 397.
46. Brand, U. and Wissen, M. (2012) 'Global environmental politics and the imperial mode of living: Articulations of state – capital relations in the multiple crisis', *Globalizations* 9(4): 547–560.
47. Dawsey, J. (2018) 'Trump on climate change: 'People like myself, we have very high levels of intelligence but we're not necessarily such believers.". *Washington Post* November 27.
48. Hook, L. and Hornby, L. (2018) 'China emerges as powerbroker in global climate talks'. *Financial Times* November 16.
49. Davidson, H. (2018) 'China on track to lead in renewables as US retreats, report says'. *The Guardian* 10 January.
50. Zhong, Y. and Hwang, W. (2015) 'Pollution, institutions and street protests in urban China', *Journal of Contemporary China* 25(98): 216–232.

51. Whitley, R. (2009) 'US capitalism: A tarnished model?', *Academy of Management Perspectives* 23(2): 11–22; Kirshner, J. (2014) *American Power after the Financial Crisis* (Ithaca: Cornell University Press).
52. Such is the growing sensitivity about this topic and ideological conformity more generally that I was asked to refrain from describing China's economy as capitalist by the editor of a prominent Chinese journal in a solicited article.
53. Liu, Y. and Chen, D. (2012) 'Why China will democratize', *The Washington Quarterly* 35(1): 41–63.
54. Tsai, K.S. (2007) *Capitalism Without Democracy: The Private Sector in Contemporary China* (Ithaca: Cornell University Press).
55. Pei, M. (2016) *China's Crony Capitalism* (Harvard University Press).
56. McGregor, R. (2010) *The Party: The Secret World of China's Communist Rulers* (New York: HarperCollins).
57. Gamble, A. (2009) 'The Western ideology', *Government and Opposition* 44(1): 1–19.
58. Baker, A. (2010) 'Restraining regulatory capture? Anglo-America, crisis politics and trajectories of change in global financial governance', *International Affairs* 86(3): 647–663.
59. Ip, G. (2018) 'A costly, deadly obsession with coal'. *Wall Street Journal* June 6; McLannahan, B. (2017) 'Goldman Sachs: Occupying Washington again'. *Financial Times*, 20 January.
60. Davidson, H. (2018) 'China on track to lead in renewables as US retreats, report says'. *The Guardian* 10 January.
61. Feng, E. (2018) 'China's annual coal consumption rises for first time in 3 years', *Financial Times*, February 28.
62. Suzuki, S. (2014) 'Journey to the West: China Debates Its 'Great Power' Identity', *Millennium - Journal of International Studies* 42(3): 632–650.
63. Norris, W.J. (2016) *Chinese Economic Statecraft: Commercial Actors, Grand Strategy, and State Control* (Ithaca: Cornell University Press).
64. Beeson, M. (2018) 'Geoeconomics with Chinese characteristics: The BRI and China's evolving grand strategy', *Economic and Political Studies* 6(3): 240–256.
65. Zhang, X. and James, K. (2017) 'From Wealth to Power: China's New Economic Statecraft', *The Washington Quarterly* 40(1): 185–203.
66. Rao, D.N. and Baruah, A.G. (2018) 'Is China going green by dumping brown on its BRI partners?'. *East Asia Forum* 25 September.
67. Chung, C.-p. (2009) 'The "good neighbour policy" in the context of China's foreign relations', *China: An International Journal* 7(1): 107–123.
68. Fawthrop, T. (2018) 'The unfolding Mekong development disaster.' *The Diplomat*, 41, April.

69. Bell, D.A. (2015) *The China Model: Political Meritocracy and the Limits of Democracy* (Princeton: Princeton University Press).
70. Beeson, M. (2018) 'Coming to terms with the authoritarian alternative: The implications and motivations of China's environmental policies', *Asia & the Pacific Policy Studies* 5(1): 34–46.
71. Asara, V., Profumi, E. and Kallis, G. (2013) 'Degrowth, democracy and autonomy', *Environmental Values* 22(2): 217–239.
72. Hausknost, D. (2017) 'Degrowth and democracy'. In C. L. Spash (Ed.), *Routledge Handbook of Ecological Economics: Nature and Society* (London: Routledge), p. 464.
73. Newell, P. (2012) *Globalization and the Environment: Capitalism, Ecology and Power* (Cambridge: Polity), p. 147 [emphasis in original].
74. Meyer, R. (2018) 'The Democratic Party wants to make climate policy exciting'. *The Atlantic*, December 5.
75. Barbier, E.B. (2010) *A Global Green New Deal: Rethinking the Economic Recovery* (Cambridge: Cambridge University Press).
76. Barbier, E.B. (2010) *A Global Green New Deal: Rethinking the Economic Recovery* (Cambridge: Cambridge University Press).
77. Beeson, M. and Bell, S. (2009) 'The G-20 and international economic governance: Hegemony, collectivism, or both?', *Global Governance* 15(1): 67–86.
78. Beeson, M. and McDonald, M. (2013) 'The politics of climate change in Australia', *Australian Journal of Politics & History* 59(3): 331–348.
79. Garnaut, R. (2010) 'Policy framework for transition to a low-carbon world economy', *Asian Economic Policy Review* 5(1): 19–33; Garnaut, R. (2011) *The Garnaut Review 2011: Australia in the Global Response to Climate Change* (Melbourne: Cambridge University Press).
80. Critical junctures are 'relatively short periods of time during which there is a substantially heightened probability that agents' choices will affect the outcome of interest.' Capoccia, G. and Kelemen, R.D. (2007) 'The study of critical junctures: Theory, narrative, and counterfactuals in historical institutionalism', *World Politics* 59(April), p. 348.
81. Some argue that human beings may be collectively developing a 'global brain' that will raise our collective consciousness. To judge by the behaviour of some populist leaders, there is plainly some distance to travel. See Breyer, C., Heinonen, S. and Ruotsalainen, J. (2017) 'New consciousness: a societal and energetic vision for rebalancing humankind within the limits of planet Earth', *Technological Forecasting and Social Change*: 1147–15.
82. Plumwood, V. (2005) *Environmental Culture: The Ecological Crisis of Reason* (London: Routledge), 238.

CHAPTER 3

The Limits to International Cooperation

The nature of Nature means doing something about environmental problems necessarily involves developing hitherto almost unimaginable levels of international cooperation. I say 'almost' because there *is* an existing institutional infrastructure of 'global governance', and there is at least one enduring example of institutionalized cooperation, albeit at the regional level. Unfortunately, the institution in question—the European Union (EU)—is at the center of a rising tide of 'bad' populism that threatens to render it ineffective at best, or to tear it apart at worst.[1] The intention of this chapter is to outline current international efforts to cooperate and solve collective action problems, especially in the area of global climate change. While the record thus far is somewhat underwhelming, perhaps the point to emphasize at the outset is that it is at least encouraging that *any* progress has been made at all. As we shall see, there are formidable barriers to cooperation, not the least of which is the widely held belief among international relations scholars that long-term cooperation between states of any kind is inherently implausible, let alone ones with radically different histories, views of the world and developmental challenges.[2]

This chapter consequently begins with a brief discussion of the way many international relations (IR) specialists think about the prospects for international cooperation. While policymakers are famously skeptical of the worth of much IR scholarship,[3] some elements of IR thinking have been influential,[4] so it is important to know what IR scholars think and why. The rest of the chapter considers why cooperation has proved so difficult and the sorts of obstacles that will have to be overcome if it is to be achieved. The good news is that there are important examples of international cooperation in a number of challenging issue areas, including climate change mitigation. The bad news is that their collective impact has been relatively modest thus far, especially given the scale of the problems and the limited time available to address them. It is for this reason that pressure from below is arguably needed to hold policymakers' feet to the fire.

Making Sense of International Cooperation

By far the most influential school of thought in IR scholarship is realism, which has exerted a direct and indirect influence on policymakers over the years.[5] Many realists argue that they are simply—as the name suggests—describing reality, not making a value judgment about how the world ought to be. The relationship between theory and practice is clearly complex and contentious, but it is striking that many prominent policymakers in the US and China have been deeply influenced by realist thinking whether they are aware of it or not. Some of them, such as former US Secretary of State Henry Kissinger and National Security Adviser Zbigniew Brzezinski, were scholar-practitioners and capable of articulating and acting on the assumptions realism makes about the world. Significantly, China's policymakers are also influenced by a long-standing tradition of realist scholarship, and that country's geopolitical trajectory seems to fulfill some of the assumptions realists make about the international order and the forces that shape it.[6] The point to emphasize is that 'the character of international life is determined by the beliefs and expectations that states have about each other, and these are constituted largely by social rather than material structures'.[7]

So what do realists think, and why? If they are correct in their assumptions about motivations of policymakers' actions, this could be a potentially fatal flaw for international cooperation. Simply put, realists think that the world is a self-help system, characterized by a competitive struggle for survival and ascendancy, in which there is no higher authority to impose order or justice. In such an environment, cooperation is inherently unlikely, or is likely to be self-serving and instrumental at best. The idea that states—the most powerful of which are still by far the most consequential actors in the international system—would put aside national interests in favor of collective or international ones is utopian wishful thinking, realists contend.[8] Looking around the world today, one might be forgiven for thinking they had a point.

Indeed, the arguments put forward by some realist scholars look all too persuasive. John Mearsheimer is probably the most prominent and influential realist scholar in the world at the moment and his critique of liberalism—the principal alternative theoretical and policy-oriented paradigm in IR—is depressingly cogent and persuasive.[9] It is also very relevant in an era when 'bad' populism is on the rise and Donald Trump is actively repudiating the sort of liberal internationalism that has defined America's international role since World War II,[10] a reality many liberals have had to recognize.[11] The idea that democracy and free market capitalism are not only normatively and practically superior, but are also sweeping all before them, looks like implausible wishful thinking at present.[12] Perhaps the most sobering aspect of Mearsheimer's critique, however, is his claim that liberalism is prone to overreaching in its efforts to remake the world in its image, the follies and hubris of George W. Bush's efforts to remake Iraq being Exhibit A in this regard.[13]

The bottom line, as far as realists generally and Mearsheimer in particular are concerned, is that 'no liberal state has ever shown serious interest in helping other states to gain economic advantages at its expense just to fight global injustice, and there is little reason to think it ever will'.[14] There is plainly something in this argument, but 'reality' is not quite as clear cut—or the future quite as bleak—as this sort of analysis claims. On the one hand, it is evident that liberals have a point when they claim that the US created an international order that benefitted others as well as themselves[15]—as even prominent 'critical' scholars have acknowledged.[16] True, the motivations for such policies may have been less benign and more grand strategic than admirers of American policies might be willing to concede,[17] but the net effect was the same. The point to emphasize is that the calculation of the national interest is often a good deal more contingent and contested than realist scholars allow.[18]

The hopeful implication of this point is that in an era when it has become increasingly evident that the domestic political and economic stability of even the most powerful and capable states is being directly threatened by potentially runaway climate change,[19] even the most nationalistic of states will be forced to deal with such *realities*. This does not mean that they will do so in a rational, constructive or cooperative way, of course, but it will inevitably force a process of adaptation. The challenge is to encourage political elites to adopt sustainable policies, rather than ad hoc authoritarian responses that are designed primarily to keep a lid on environmentally induced unrest.[20] This is where international cooperation, especially the institutionalized variety, is potentially so important. Before considering in any detail how international institutions have operated, however, it is important to say something about the role of states as they are potentially the biggest obstacles to, as well as the indispensable fulcrum of, inter*national* cooperation.

STATES: PART OF THE SOLUTION?

The modern state hasn't been around for that long. If there is one date students of IR are likely to remember from their studies, it is 1648, when the signing of the Treaty of Westphalia brought an end to the Thirty Years' War and simultaneously enshrined the notion of national sovereignty.[21] Before this time, various other forms of political organization could be found around the world, but the state enjoyed decisive advantages that meant that no other form of political organization could hope to survive, let alone compete.[22] Not the least of these advantages was the ability to inflict military defeat on potential rivals.[23] Little wonder that many realists still think the potentially brutal nature of the inter-state system determines and by implication *limits* the sorts of policies state policymakers can pursue.[24] No matter how enlightened and progressive individual leaders may be, the nature of the system and the behavior of other states compel them to privilege national interests and security—or so the argument goes.

At first blush, European history seems to confirm this thesis. Not only was the modern state invented in Western Europe, but its initial historical development owed much to the specific contingencies of the Europe's geography and history. State survival was dependent on developing key capacities and institutions, especially in the face of technological and organizational advantages that threatened to give decisive first mover advantages to the states the developed them.[25] In this context, no innovation has proved to be more consequential that adding 'nation' to state, as became so commonplace throughout Europe in the nineteenth century.[26] Napoleon Bonaparte's *Grande Armée* benefitted from the introduction of mass conscription and the inculcation of national values, something other states had to replicate if they hoped to compete. The rise of France as a disruptive, revolutionary power also led to the development of an effective balance of power, epitomized by the Concert of Europe, which brought an unprecedented degree of stability to the world's most belligerent continent.[27] Even at this point in history, it should be noted that security threats could actually generate surprisingly cooperative behavior, especially at a time when the environment is currently seen as a major security threat.[28]

The emergence and consolidation of the nation-state throughout Europe also enabled European powers to dominate and colonize other parts of the word. In the process, the nation-state was exported around the world, albeit with radically different levels of success and capability.[29] Indeed, so ubiquitous did the nation-state become and so prized was the autonomy it promised, that national sovereignty was enshrined in the founding principles of the United Nations. The point of this short historical digression is to emphasize just how important nationalism and the primacy of the state have been in both Europe's rise, and in the large number of new nation-states that were created in the aftermath of decolonization. While many of these new states may not have achieved anything like the degree of power, authority and capacity enjoyed by many of the states in what we now call the 'global North',[30] this did not make them any the less concerned about possible infringements of national sovereignty. On the contrary, many states in the South are more protective of state sovereignty than their more established counterparts in the North.[31]

Even if observers such as Stephen Krasner are correct in claiming that national interests trump any consideration of international rules and norms in the calculations of policymakers, he also makes the important point that 'an environment devoid of institutional structures would be akin to a state of nature in which behavior is driven only by short-term calculations of interest and action constrained only by the power of others.'[32] In other words, it would something like the anarchical conception that is central to realist thinking, and which suggests that international cooperation is unlikely and unreliable. But it's not: the reality is that institutions *do* exist and this has undoubtedly influenced the behavior of states as a consequence. Indeed, there is a long-standing argument in IR scholarship that states actually *demand* the creation of international institutions as they directly benefit from the resolution of collective action problems.[33] Consequently, we do not live in an ungoverned state of nature, even if some of the institutions of global governance are not as effective, democratic or accountable as we might like them to be.[34]

This is why—I think—that the example of the EU is so important and even mildly inspiring: the EU demonstrates that not only have nations been willing to 'pool' sovereignty in ways that realists consider improbable, but the EU has also pacified one of the most violent regions on the entire planet.[35] This was after all, the continent that gave us the 100 Years' War, religious persecution on an epic scale, industrialized genocide, World Wars I and II and countless other horrors. Yet the EU has also pioneered new transnational forms of environmental cooperation and policy experimentation that have proved surprisingly effective at times.[36] As we shall see, however, it is also at the forefront of a resurgence of virulent far-right and nationalistic politics that threatens to tear the organization apart or—more likely—reduce the EU to a pale imitation of its former self.[37] This is not a good omen for those of us hoping to see more rather than less institutionalized cooperation. It is a possibility that we have to take seriously, however.

WHY IS INTERNATIONAL COOPERATION SO DIFFICULT?

One of the reasons that international cooperation is so difficult is that its principal institutional manifestations are overwhelmingly associated with, and a product of, what we might best describe as American hegemony.[38] Although many scholars, especially in the US, prefer to talk about leadership,[39] the breadth, depth and institutionalized nature of American dominance has given it an influence no other state has been able to match.[40] Even Britain in its imperial heyday was not the sort the sort of global power that the US became in the aftermath of World War II when its economic and strategic position became increasingly unparalleled, especially following the unexpected demise of the Soviet Union.[41] Consequently, and despite the fact that American hegemony provided an environment in which many potential rivals such as China actually prospered,[42] many states in the South view America's influence and dominance with great suspicion.[43]

Yet major differences of opinion about how the world should be governed were a problem even before climate change became the quintessential collective action problem of the era. There was—and still is—a sense that the world the US helped to make reflected values and principles that reflected and *advantaged* American interests, rather than those of the much invoked but seldom seen 'international community'.[44] The essence of this sort of 'counter-hegemonic' impulse is captured by Robert Kuttner, who argues that

> [t]oday, it is not communists, but capitalists, who are seeking to impose a single economic model on the globe. On balance, the institutions of global

governance tend to reflect and reinforce rather than challenge that dominance. Globalism has been great at advancing the interests of capital and feeble at defending or enlarging the domain of human rights. The home of democracy – or antidemocracy – continues to be the national polity.[45]

Whatever one may think about the historical role of the US, it has clearly helped to anchor a particular international order that is currently being challenged and/or undermined by a number of long-term trends: the possible decline of American power itself, the rise of China and other powers, and the declining confidence in the extant order, epitomized by the emergence of Donald Trump and his disdain for multilateralism in particular and the value of international cooperation more generally.[46] In such circumstances, and given America's withdrawal from the Paris Agreement on climate change mitigation, the prospects for meaningful cooperation do not look good. Indeed, in such circumstances the pessimists and the realists look to have some compelling arguments, which need to be taken seriously and—if possible—repudiated. Given that skeptics have little to offer in the way of long-term solutions, possible remedies are worth taking seriously; even some of the more implausible and hitherto unlikely ideas actually begin to look attractive. What, after all, is the alternative?

One surprising glimmer of hope may be found in the fact that some of the most unlikely people now not only take climate change seriously, but are thinking about it differently and actively preparing for its consequences.[47] This potential good news needs to be tempered by the fact that people who are professionally preoccupied with conventional security tend to see things through a fairly narrow window of 'strategic concerns', in which environmental sustainability may not be foremost.[48] On the contrary, many of the most plausible scenarios mapped out by strategic thinkers in places like the Pentagon work on the assumption that things are likely to get much worse as a consequence of climate change and that it is their job to prepare possible military responses.[49] One of the more significant conclusions of a major report prepared by the Center for Strategic and International Studies (CSIS) and the Center for New American Security (CNAS) was that in the event of major climate change induced social crises, 'among political systems, authoritarian ideologies would certainly be the "winners." One way or the other, severe climate change will weaken the capacity of liberal democratic systems to maintain public confidence'.[50]

While this conclusion may be admirably 'realistic' and nonideological, it is also noticeably at odds with the traditional conventional wisdom about the merits and durability of democracies. Significantly, this report was produced well before the current upsurge in populism and the global financial crisis (GFC) of 2008, which did so much to undermine confidence in the American model of politics and economics.[51] One of the principal reasons conventional security analysts have become so preoccupied with climate change is not necessarily because of their commitment to democracy, but because global warming threatens to directly affect more traditional questions of national security. There is a good deal of all-too-plausible evidence and research directly linking deteriorating environment conditions with the outbreak of conflict.[52] The possibility that this will translate into urgent action as a consequence, however, is far from clear: most of the conflict takes the form of civil wars and—despite some sobering claims about the climate's role in historical conflicts[53]—they have tended to be confined to some of the least strategically important areas of the world.[54]

And yet the idea that the world can be divided into zones of peace and conflict,[55] in which the North insulates itself from the problems of the South, through its superior military capabilities if necessary, looks increasingly implausible. On the one hand, it has been persuasively argued that crises such as the one that has plagued Syria for nearly a decade and which has destabilized the entire Middle East had environmental origins, rather than geopolitical ones.[56] On the other hand, no country can insulate itself from the impact of rising temperatures: Whether it is increased wildfires and 'weather events' in the US and Australia, or chronic water shortages in China, even the most modern and lavishly equipped military is no defense against a changing natural environment. This is not to say that such military capabilities will not be deployed or employed to pursue other national goals once established, of course. On the contrary, as climate change sows further political discord, the chances of strategic miscalculation can only grow. The case for effective action and cooperation would seem overwhelming, therefore. The question is, even if 'serious', hard-headed strategic thinkers are now concerned about climate change, can anything useful and environmentally sustainable be achieved?

Coming Up with a Common Agenda

The scientific community has alerted us all to the nature of climate change and its possible impacts. As we have seen, many people have chosen to ignore or dismiss this message for one reason or another, but there are plenty of people who do take the scientific warnings seriously. Many of them are powerful and influential, and some of them are even members of the business community—or the capitalist class, if you prefer. Quite how you prefer to describe the relatively small group of people who control key investment decisions and who largely determine the way we collectively organize the production and distribution of desirable goods probably reflects whether you think business is part of the solution or the cause of many of the problems we collectively face. Given the pervasive and diverse nature of business (or capital), though, it's probably both. Ultimately, captains of industry are no more likely to survive and prosper than the rest of us. Some of them plainly get this, which is why they are buying postapocalyptic bolt-holes in New Zealand, no doubt.[57] Significantly, even prominent members of the American ruling class, such as former vice president Al Gore, recognize that 'the radically elevated role of money in politics has given the forces representing wealth and corporate power sufficient strength to advance their agenda even when a sizeable majority of the American people oppose it.[58]

The US is in many ways a special case, even 'exceptional',[59] given its still uniquely powerful position. Without American or Chinese cooperation—given the latter's growing influence and contribution to many of the world's collective environmental problems—nothing of significance is likely to occur. This is why the Trump administration's repudiation of international cooperation and climate change science is so significant and frankly depressing.[60] China's entirely unconstructive role in the abortive climate change summit in Copenhagen is also indicative of its ability to stymie agreement and frustrate efforts to develop a cooperative agenda—even when the US, in the form of former president Barack Obama is seemingly on board and lending support to a collective response.[61] In such circumstances, one might be forgiven for thinking that all is lost and meaningful collective action is simply too difficult. Perhaps it is, but there are some grounds for optimism, not the least of which is the fact that states feel obliged to take part in the sort of 'regime complexes' that—in theory, at least—offer some scope for policy flexibility and the inclusion of a wider array of actors.[62]

Indeed, in many ways, managing climate change is the definitive test of global governance, or 'the conscious, goal-oriented collective actions of state and non-state actors to develop new responses to problems that are both transnational and beyond the capacity of individual governments or organizations to resolve'.[63] Many observers hold out great hopes about the ability of non-state actors to 'help states gain the technology, expertise and confidence to formulate and implement more ambitious contributions, and to build interest group support to pursue them.'[64] There is no doubt that global governance *has* become a more complex process that involves a much wider array of actors than simply individual nation-states. States, however, still remain the most consequential actors when it comes to actually implementing policy. As Drezner points out, for all the increased prominence of intergovernmental organizations (IGOs), their 'strong monitoring and enforcement capabilities are used for regulatory matters only when a great powers consensus exists on the content of the regulations to be enforced'.[65]

Two points are worth emphasizing about the enduring importance of states, therefore. First, without the agreement of the most powerful states in the system, not much is likely to happen of any consequence as far as climate change or any other issue is concerned. This is why China and the US, their bilateral relationship, and their capacity for leadership are so potentially important. The second point to note is that states do not make policy in a politically neutral or technocratic environment. On the contrary, state leaders have one eye on their respective and potentially very different domestic audiences and national circumstances,[66] a reality that helps to explain the divergence of positions that have emerged between the North and South in climate negotiations.[67] The idea that some countries might have made a greater contribution to common climate problems and that they might consequently have a greater role to play in addressing them has been captured in the idea of 'common but differentiated responsibilities'.[68] These sorts of policy innovations have helped to overcome entrenched differences and potential divisions, and allowed states to negotiate more than 1000 multilateral, and 1500 bilateral environmental agreements, a number that continues to grow. But as Christoff and Eckersley point out, however, 'the paradox of environmental multilateralism is that while environmental treaty output…has increased, the most serious and irreversible global environmental problems facing the international community have worsened'.[69]

Scaling Down?

Even though a number of UN-sponsored bodies and initiatives such as the Intergovernmental Panel on Climate Change (IPCC) and the UN Framework Convention on Climate Change (UNFCCC) have done important work in detailing the impact of climate change and providing a potential forum for its mitigation, only modest progress has been made. To be sure, there has been a number of high-profile, well-intentioned summits in Kyoto, Rio de Janeiro, Copenhagen and most recently Paris, but their ability to translate worthy rhetoric into meaningful action has been limited. The 'Paris Agreement' of 2015 appeared to represent a major step forward following the dashed expectations of Copenhagen: 175 countries signed the final agreement designed to limit future global temperature increases to 2 °C. But the Trump administration's decision to withdraw from the agreement was both predictable and potentially fatal for the credibility of the agreement as a whole.[70] Other countries, such as Australia, which also has many opponents of multilateral environmental action, can claim that their efforts will be meaningless without the participation of the greatest contributors to the problem and that they should continue to support major export industries, such as coal.[71]

At one level, of course, it is perfectly correct to argue, as conservative figures repeatedly do,[72] that Australia's actions alone will have little impact on global CO_2 emissions. Australia's total emissions are a small fraction of the world's, largely as a consequence of its relatively small population. On a per capita basis, however, Australia's emissions are among the world's highest and continue to grow.[73] Yet from a 'national interest' perspective, policymakers may reasonably ask whether the short-term sacrifices in terms of possible job losses and the balance trade are worthwhile. This is an especially salient consideration in Australia where the electoral cycle is especially truncated[74] and where there is a growing divide between urban and regional voters about the importance of the environment relative to 'development'.[75] In this regard, the Australian experience highlights more general problems about conflicting domestic priorities and possible international obligations.

Australia also illustrates the difficulties and political dangers of trying to provide leadership on issues as complex as climate change. Although no continent is likely to be more adversely affected by global warming than Australia—already the driest continent in the world—any politician trying to demonstrate leadership on this issue does so at his or her peril. Australia's notorious revolving door prime ministership—seven in the last ten years— has been exacerbated by the question of how to deal with climate change. Despite some well-intentioned efforts from former Prime Ministers Kevin Rudd and Malcolm Turnbull in particular, both of them lost their jobs in part because of their failure to deliver on environmental policies.[76] Promoting 'good international citizenship' in a frequently parochial political environment in which the coal lobby has traditionally loomed large has proved beyond any Australian leader thus far. But as Henry Shue points out, international leadership has to come from somewhere,[77] especially in the absence of the American variety. If a country such as Australia cannot set a good example, given its relative fortunate circumstances, it is worth asking who can.[78]

The rather sobering reality is that for all the policy innovation, the significant contributions of various non-state actors, and the creation of potentially important institutions through which to disseminate collective initiatives and expertise,[79] the major powers continue to focus on 'securing their own national interest and avoiding costly commitments to emission reductions or long-term funding for adaptation'.[80] Indeed, Green and Colgan conclude that in the area of environmental politics, at least, 'contrary to some claims of a "power shift" from states to different nonstarter actors, our findings do not suggest that there is a devolution of power away from the state'.[81] Adding to this problem is the fact that between states there are still major and enduring disparities of power and influence that mean that some states—and the people they purport to represent—have much less ability to shape policy than others. As Stevenson and Dryzek note, 'the formal equality of all states to participate in the UNFCCC negotiations is highly compromised in practice. To compound the problem, effective inclusion in negotiation does not guarantee effective inclusion in deliberation'.[82]

There is, however, at least one area where effective action appears to be occurring, despite—or more accurately, *because* of—the existence of climate change denying leaders such as Donald Trump. In the US in particular, local governments have been pioneering the adoption of sustainable public policies that are specifically, and successfully, designed to reduce Greenhouse gas emissions by 'greening' local building codes, creating commercial waste reduction programs, encouraging water conservation and promoting a range of other conservation and energy reduction initiatives.[83] Important and effective initiatives have been developed via the UN's World Congress of Local Governments for a Sustainable Future, which have helped pass on ideas about environmental best practice at the local level, even without the encouragement or even opposition of central governments. Given that California on its own is reckoned to be the world's fifth biggest economy, the efforts of such 'local' actors are not inconsiderable. Nor are they confined to the US. Across Germany, Scandinavia and a total of over 40 other countries, similar initiatives are being undertaken at the local level.[84] As Schreurs points out, 'there are many specific examples of local governments moving faster than and influencing central and national governmental decisions. This leads to the important conclusion that local government action can have powerful effects far beyond local jurisdictions'.[85]

While this may be true of Western democracies, however, the picture in authoritarian China is more complex and contradictory. On the positive side, it is clear that the central government is alert to the growing unhappiness of China's expanding environmentally conscious middle class population in major urban centers. It is also widely noted that China is the world's largest in renewable energy and that it is making significant efforts to transform the structure of an economy that has generated massive pollution as well as great wealth.[86] There are, however, limits to what China's top down, command and control system can achieve given the sheer scale of the challenges China faces and the difficulty of getting local elites and provincial governments to cooperate.[87] In China's case, and in striking contrast to the US, for example, the local governments and their truncated time horizons and parochial growth targets are actually part of the problem. Kevin Lo sums up the problem this way:

> The lack of instruments to enforce the authoritarian policy regime and the failure to set priorities between economic growth and climate protection have resulted in an inadequate setting of incentives and disincentives for local business managers and representatives of local governments, creating a situation of de facto neoliberal environmentalism. In other words, local governments and businesses enjoy a surprisingly high degree of freedom and flexibility to manage their own energy consumption in spite of the overt authoritarian rule.[88]

The inescapable, rather deflating conclusion is that, in China's nondemocratic regime, at least, local government and participation is no guarantee of good environmental outcomes; quite the contrary in fact. Indeed, local governments may actually encourage just the sort of 'race to the bottom' in lax environmental standards that many feared in an era when businesses are able to exploit local conditions—even when those conditions are found in a supposedly noncapitalist system.[89] Nevertheless, the growth of environmental NGOs and the emergence, albeit under increasingly constricted circumstances,[90] of a more influential civil society do suggest that even in China, an informed and active citizenry can place pressure on governments at all levels to enact more sustainable policies.[91]

Conclusion

One conclusion that can be uncontroversially drawn from the overall discussion is that the international institutional architecture that has been created to encourage and facilitate cooperation over the last half century or so is not working as it once did. To be sure, it may never have fulfilled the hopes of some of its architects, and may always have served some interests more than others, but whatever view you may have of its value and importance, the international order is not what it used to be.[92] In some ways, this may not be an unambiguously bad thing, of course: critics may welcome the demise of a system that they claim furthered the interests of the rich and powerful at the expense of the poor and marginalized. But even if that is correct, *any* order may be better than no order; after all, imperfect as it may have been, the so-called liberal international order created largely under the auspices of American hegemony was directly associated with the reconstruction of Western Europe and the unprecedented rise of Asia—of which China is but the latest installment—not a bad outcome by world historical standards.

At a historical moment when the US in the shape of the Trump administration seems to have given up on its former leadership role, and China looks far from able (or willing) to replace it,[93] major unanswered questions hang in the air about quite how international relations generally are going to be conducted in the next few decades. This is an interesting academic question, to which there are already a number of possible answers.[94] Unfortunately, however, and at the risk of unforgivable hyperbole, this may well be the most important decade or so in the history of the human race. Yes, that is a grand claim but one supported by a growing body of evidence that suggests that we have a very limited window of opportunity to turn our collective destiny around, before we consign future generations to a nightmarish, Darwinian struggle for survival, from which it is far from certain they will actually emerge. As Steffen et al. rather dispassionately observe:

> social and technological trends and decisions occurring over the next decade or two could significantly influence the trajectory of the Earth System for tens to hundreds of thousands of years and potentially lead to conditions that resemble planetary states that were last seen several millions of years ago, conditions that would be inhospitable to current human societies and to many other contemporary species.[95]

If that's not an issue worth getting agitated and active about, it's hard to know what is.

Notes

1. Interestingly, global warming may be contributing to this. See *The Economist* (2018) 'A hot summer is giving Europe's populists a boost', July 26.
2. See, Booth, K. (ed. 2011) *Realism and World Politics* (London: Routledge).
3. Reus-Smit, C. (2012) 'International relations, irrelevant? Don't blame theory', *Millennium – Journal of International Studies* 40(3): 525–540.
4. John Maynard Keynes' observations about the enduring power of some ideas is worth remembering: 'The ideas of economists and political philosophers both when they are right and when they are wrong, are more powerful than is commonly understood. Indeed the world is ruled by little else. Practical men, who believe themselves to be quite exempt from any intellectual influence, are usually the slaves of some defunct economist. Madmen in authority, who hear voices in the air, are distilling their frenzy from some academic scribbler of a few years back.'
5. Hans Morgenthau, for example, exerted an enormous influence over both the study and practice of IR in the US in the postwar period. See Gaskarth, J., ed. 2018. *Hans J. Morgenthau and the American Experience* (Oxford: Oxford University Press); Morgenthau, H.J. (1972 [1948]) *Politics Among Nations: The Struggle for Power and Peace*, 5th ed. (New York: Knopf).
6. Johnston, A.I. (1995) *Cultural Realism: Strategic Culture and Grand Strategy in Chinese History* (Princeton: Princeton University Press).
7. The quote is from Alexander Wendt, who is the most prominent exponent of a 'constructivist' approach to IR that emphasizes the social construction of reality. Wendt, A. (1999) *Social Theory of International Politics* (Cambridge: Cambridge University Press), p. 20.
8. Mearsheimer, J.J. (2001) *The Tragedy of Great Power Politics* (New York: W.W. Norton); Walt, S.M. (2018) 'US grand strategy after the Cold War: Can realism explain it? Should realism guide it?', *International Relations* 32(1): 3–22.
9. Mearsheimer, J.J. (2018) *The Great Delusion: Liberal Dreams and International Realities* (New Haven: Yale University Press).
10. Some observers argue that Trump's foreign policies are essentially realist, not populist, and welcome for that reason. See, Schweller, R. (2018) 'Three cheers for Trump's foreign policy', *Foreign Affairs* (September/October).
11. Haas, R.N. (2018) 'Liberal world order, R.I.P.'. *Project Syndicate* March 21.
12. Luce, E. (2017) *The Retreat of Western Liberalism* (New York: Atlantic Monthly Press).
13. Packer, G. (2005) *The Assassin's Gate: America in Iraq* (New York: Farrar, Strauss and Giroux).
14. Mearsheimer, J.J. (2018) *op cit*, pp. 128–29.

15. Ikenberry, G.J. (2011) *Liberal Leviathan: The Origins, Crisis, and Transformation of the American World Order* (Princeton: Princeton University Press).
16. Cox, R.W. (1987) *Production, Power, and World Order: Social Forces in the Making of History* (New York: Columbia University Press).
17. Beeson, M. and Watson, N. (2019) 'Is international leadership changing hands or disappearing? China and the USA in comparative perspective', *Asian Perspective*.
18. Finnemore, M. (1996) *National Interests in International Society* (Ithaca: Cornell University Press).
19. Stern, N. (2008) 'The economics of climate change', *American Economic Review* 98(2): 1–37.
20. Beeson, M. (2010) 'The coming of environmental authoritarianism', *Environmental Politics* 19(2): 276–294.
21. The story of the state's rise is more complex than this depiction might suggest, but beyond the scope of this discussion. See Teschke, B. (2003) *The Myth of 1648: Class, Geopolitics, and the Making of Modern International Relations* (London: Verso).
22. Spruyt, H. (2002) 'The origins, development, and possible decline of the modern state', *Annual Review of Political Science* 5127–149; Buzan, B. and Little, R. (2000) *International Systems in World History: Remaking the Study of International Relations* (Oxford: Oxford University Press).
23. Tilly, C. (1985) 'War making and state making as organized crime', in, P.B. Evans and et al. *Bringing the State Back In* (Cambridge University Press): 169–191.
24. Waltz, K.N. (1979) *Theory of International Politics* (New York: McGraw-Hill).
25. Fukuyama, F. (2011) *The Origins of Political Order: From Prehuman Times to the French Revolution* (New York: Farrar, Strauss and Giroux).
26. Hobsbawm, E.J. (2012) *Nations and Nationalism Since 1780: Programme, Myth, Reality* (Cambridge: Cambridge University Press).
27. Mitzen, J. (2013) *Power in Concert: The Nineteenth-century Origins of Global Governance* (University of Chicago Press).
28. Dyer, G. (2010) *Climate Wars: The Fight for Survival as the World Overheats* (Oxford: Oneworld); Dalby, S. (2014) 'Rethinking Geopolitics: Climate Security in the Anthropocene', *Global Policy* 5(1): 1–9.
29. Watson, A. (1994) 'European international society and its expansion', in, H. Bull and A. Watson *The Expansion of International Society* (Oxford: Clarendon Press): 13–32. Cf., Dunne, T. and Reus-Smit, C. (eds. 2017) *The Globalization of International Society* (Oxford: Oxford University Press).
30. Jackson, R. (1990) *Quasi-states: Sovereignty, International Relations and the Third World* (Cambridge University Press).

31. The newish states of Southeast Asia illustrate many of these issues, especially the reluctance to compromise national sovereignty in pursuit of some overarching collective good. See, Beeson, M. (2003) 'Sovereignty under siege: globalisation and the state in Southeast Asia', *Third World Quarterly* 24(2): 357–374.
32. Krasner, S.D. (1999) *Sovereignty: Organized Hypocrisy* (Princeton: Princeton University Press), p. 57.
33. Keohane, R.O. (1982) 'The demand for international regimes', *International Organization* 36(25): 325–355.
34. Weiss, T.G. (2013) *Global Governance: Why? What? Whither?* (Cambridge: Polity).
35. Gleditsch, N.P. (1995) 'Democracy and the future of European peace', *European Journal of International Relations* 1(4): 539–571.
36. Quahe, S. (2018) 'EU in crisis: what implications for climate and energy policy?', *Asia Europe Journal* 16(2): 169–182; Sabel, C.F. and Zeitlin, J. (2010) *Experimentalist Governance in the European Union: Towards a New Architecture* (Oxford: Oxford University Press).
37. Kreuder-Sonnen, C. (2018) 'An authoritarian turn in Europe and European Studies?', *Journal of European Public Policy* 25(3): 452–464.
38. Agnew, J. (2005) *Hegemony: The New Shape of Global Power* (Philadelphia: Temple University Press).
39. Nye, J.S. (1990) *Bound to Lead: The Changing Nature of American Power* (New York: Basic Books).
40. Kagan, R. (2012) *The World America Made* (New York: Alfred A. Knopf); Ferguson, N. (2004) *Colossus: The Price of America's Empire* (New York: Penguin).
41. Wohlforth, W.C. (1999) 'The stability of a unipolar world', *International Security* 24(1): 5–41.
42. Stubbs, R. (2005) *Rethinking Asia's Economic Miracle* (Basingstoke: Palgrave).
43. Kupchan, C.A. (2014) 'The normative foundations of hegemony and the coming challenge to Pax Americana', *Security Studies* 23(4): 219–257; Root, H.L. (2008) *Alliance Curse: How America Lost the Third World* (Washington: Brookings Institute Press); Saull, R. (2012) 'Rethinking hegemony: Uneven development, historical blocs, and the world economic crisis', *International Studies Quarterly* 56(2): 323–338.
44. Ellis, D., C. (2009) 'On the possibility of "international community"', *International Studies Review* 11(1): 1–26.
45. Kuttner, R. (2018) *Can Democracy Survive Capitalism?* (New York: W.W. Norton), p. 257.
46. Kagan, R. (2018) *The Jungle Grows Back: American and Our Imperiled World* (New York: Knopf).

47. McDonald, M. (2012) *Security, The Environment and Emancipation: Contestation Over Environmental Change* (London: Routledge).
48. Klein, B.S. (1988) 'Hegemony and Strategic Culture: American Power Projection and Alliance Defence Politics', *Review of International Studies* 14(2): 133–148.
49. Board of the C.M.A. (2014) *National Security and the Accelerating Risks of Climate Change* (Alexandria, VA.: CNA Corporation).
50. Campbell, K.M., Gulledge, J., McNeill, J.R., Podesta, J., Ogden, P., Fuerth, L., Woolsey, R.J., Lennon, A.T., Smith, J. and Weitz, R. 2007. *The Age of Consequences: The Foreign Policy and National Security Implications of Global Climate Change* (Washington: SSIS/CNAS), 77.
51. Kirshner, J. (2014) *American Power after the Financial Crisis* (Ithaca: Cornell University Press).
52. Homer-Dixon, T.F. (1994) 'Environmental Scarcities and Violent Conflict: Evidence from Cases', *International Security* 19(1): 5–40; Parenti, C. (2011) *Tropic of Chaos: Climate Change and the New Geography of Violence* (New York: Nation Books); Stalley, P. (2003) 'Environmental scarcity and international conflict', *Conflict Management and Peace Science* 20(2): 33–58.
53. Zhang, D.D., Lee, H.F., Wang, C., Li, B., Pei, Q., Zhang, J. and An, Y. (2011) 'The causality analysis of climate change and large-scale human crisis', *Proceedings of the National Academy of Sciences* 1–6.
54. Goldstone, Jack A. (2002) Population and security: How demographic change can lead to violent conflict', *Journal of International Affairs*, 56 (1): 3–22.
55. Singer, A. and Wildavsky, M. (1993) *The Real World Order: Zones of Peace/Zones of Turmoil* (New Jersey: Chatam House Publishers).
56. Gleick, P.H. (2014) 'Water, drought, climate change, and conflict in Syria', *Weather, Climate, and Society* 6(3): 331–340.
57. O'Connell, M. (2018) 'Why Silicon Valley billionaires are prepping for the apocalypse in New Zealand'. *The Guardian* 15 February.
58. Gore, A. (2013) *The Future* (London: W.H. Allen), 122.
59. A belief in American exceptionalism has been one of the defining features of US domestic and especially foreign policy. Unfortunately, China currently labours under the same illusion. See, Zhang, F. (2013) 'The rise of Chinese exceptionalism in international relations', *European Journal of International Relations* 19(2): 305–328.
60. Beeson, M. (forthcoming) 'Donald Trump and post-Pivot Asia: The implications of a 'transactional' approach to foreign policy', *Asian Studies Review*.
61. Conrad, B. (2012) 'China in Copenhagen: Reconciling the "Beijing climate revolution" and the "Copenhagen climate obstinacy"', *The China Quarterly* 210, 435–455.

62. Keohane, R.O. and Victor, D.G. (2011) 'The regime complex for climate change', *Perspectives on Politics* 9(01): 7–23.
63. Beeson, M. (2019) *Rethinking Global Governance* (Basingstoke: Palgrave), p. 1.
64. Chan, Sander, Harro van Asselt, Thomas Hale, Kenneth W. Abbott, Marianne Beisheim, Matthew Hoffmann, Brendan Guy, Niklas Höhne, Angel Hsu, Philipp Pattberg, Pieter Pauw, Céline Ramstein, and Oscar Widerberg. 2015. "Reinvigorating International Climate Policy: A Comprehensive Framework for Effective Nonstate Action." *Global Policy* 6(4), p. 467.
65. Drezner, D.W. (2007) *All Politics is Global: Explaining International Regulatory Regimes* (Princeton: Princeton University Press), p. 211.
66. Robert Putnam famously described this situation and the contradictions it poetically raised as a 'two level game'. See, Putnam, R.D. (1988) 'Diplomacy and domestic politics: the logic of two-level games', *International Organization* 42(3): 427–460.
67. Hurrell, A. and Sengupta, S. (2012) 'Emerging powers, North–South relations and global climate politics', *International Affairs* 88(3): 463–484.
68. Stone, C.D. (2004) 'Common but differentiated responsibilities in international law', *American Journal of International Law* 98(2): 276–301.
69. Christoff, P. and Eckersley, R. (2013) *Globalization and the Environment* (Lanham: Rowman & Littlefield), p. 168.
70. Urpelainen, J. (2017) 'Trump's withdrawal from the Paris agreement means other countries will spend less to fight climate change', *Washington Post*, November 21.
71. Karp, P. (2018) 'Australian government backs coal in defiance of IPCC climate warning'. *The Guardian* 8 October.
72. Abbott, T. (2018) 'Knowing what we do now, the Paris climate deal is passe'. *The Australian* July 4.
73. Jericho, G. (2018) 'Australia's emissions are rising. It's time for this government to quit pretending'. *The Guardian* 8 Jan.
74. Australian voters are compelled to go to the polls at least once every three years for federal elections. The idea that the long-term future of the nation let alone the planet will be a policy priority is predictably low as a consequence.
75. Ferguson, J. (2019) 'City greens versus bush mining jobs is war: Mundine'. *The Australian* January 7.
76. Sengupta, S. (2018) 'Climate change policy toppled Australia's leader. Here's what it means for others'. *New York Times* August 24; Beeson, M. and McDonald, M. (2013) 'The politics of climate change in Australia', *Australian Journal of Politics & History* 59(3): 331–348.
77. Shue, H. (2011) 'Face reality? After you! A call for leadership on climate change', *Ethics & International Affairs* 25(01): 17–26.

78. Beeson, M. (2011) 'Can Australia save the world? The limits and possibilities of middle power diplomacy', *Australian Journal of International Affairs* 65(5): 563–577.
79. Ovodenko, A. and Keohane, R.O. (2012) 'Institutional diffusion in international environmental affairs', *International Affairs* 88(3): 523–541.
80. Falkner, R., Stephan, H. and Vogler, J. (2010) 'International climate policy after Copenhagen: Towards a 'building blocks' approach', *Global Policy* 1(3), p. 256.
81. Green, J.F. and Colgan, J. (2013) 'Protecting sovereignty, protecting the planet: State delegation to international organizations and private actors in environmental politics', *Governance* 26(3), p. 494.
82. Stevenson, H. and Dryzek, J.S. (2014) *Democratizing Global Climate Governance* (Cambridge: Cambridge University Press), p. 77.
83. Linstroth, T. and Bell, R. (2007) *Local Action: The New Paradigm in Climate Change Policy* (Burlington: University of Vermont Press).
84. Rosenzweig, C., Solecki, W., Hammer, S.A. and Mehrotra, S. (2010) 'Cities lead the way in climate–change action', *Nature* 467(7318): 909.
85. Schreurs, M.A. (2008) 'From the bottom up: local and subnational climate change politics', *The Journal of Environment & Development* 17(4), p. 353.
86. Mathews, John A, and Hao Tan. 2014. "Manufacture Renewables to Build Energy Security." *Nature* 513: 166–68.
87. Kostka, Genia. 2015. "Command without control: The case of China's Environmental target system." *Regulation and Governance* (December 2014): 58–74; Eaton, Sarah, and Genia Kostka. 2014. "Authoritarian environmentalism undermined? Local leaders' time horizons and environmental policy implementation in China." *The China Quarterly* 218: 359–80.
88. Lo, Kevin. 2015. "How authoritarian is the environmental governance of China?" *Environmental Science & Policy*, 52, p. 158.
89. Qi, Ye, and Lingyun Zhang. 2014. "Local environmental enforcement constrained by central-local relations in China." *Environmental Policy and Governance* 24(3): 204–15.
90. Teets, J.C. (2014) *Civil Society Under Authoritarianism: The China Model* (Cambridge University Press).
91. Steinhardt, H Christoph, and Fengshi Wu. (2016) 'In the name of the public: Environmental protest and the changing landscape of popular contention in China.' (75): 61–82.
92. Haass, R. (2018) 'How a World order ends and what comes in its wake', *Foreign Affairs* (January/February).
93. Beeson, M. (2013) 'Can China lead?', *Third World Quarterly* 34(2): 235–252.

94. Among a rapidly expanding genre, see Acharya, A. (2014) *The End of the American World Order* (Cambridge: Polity Press); Bremmer, I. (2012) *Every Nation for Itself: Winners and Losers in a G-Zero World* (New York: Portfolio/Penguin); Luce, E. (2012) *Time to Start Thinking: America in the Age of Descent* (New York: Atlantic Monthly Press); Zakaria, F. (2008) *The Post-American World* (New York: W.W. Norton).
95. Steffen, W., Rockström, J., Richardson, K., Lenton, T.M., Folke, C., Liverman, D. and Schellnhuber, H. (2018) 'Trajectories of the Earth System in the Anthropocene', *Proceedings of the National Academy of Sciences* 115(33), p. 8253.

CHAPTER 4

Debating Populism

Populism is popular. Whether it's the remarkable increase in the number of people described as populists, or the number of academics and other commentators trying to understand the possible significance of populism, there's a lot of it about. Given that the theory and—more importantly—practice of populism is not a new phenomenon, one obvious question to consider is why populism is attracting such attention at this particular moment in history. One possible reason, I shall suggest, is that the 'populist style'[1] has spread beyond some of its more familiar breeding grounds in Latin America, Africa and Asia, and begun to appear—or, more accurately, *re*appear—in Europe and North America. Or it has if you think that Donald Trump is some form of populist leader, at least.

As we shall see, deciding what populism actually is and who really qualifies is more complex than we might expect. At one level, this is, perhaps, an expression of academic nit-picking and/or conceptual clarification of a sort that distinguishes much scholarly endeavor. At another level, however, it highlights a potential limitation of the entire debate: if we can't say with any certainty whether Trump and the leader of the British Labour Party, Jeremey Corbyn, are populists, how useful is the concept? Likewise, is the formerly wildly popular Emmanuel Macron, who came from nowhere to seize the French presidency, a populist, or are the so-called *gilets jaunes* (yellow vests) protestors an expression of a more genuine upsurge of popular sentiment and outrage?

While this chapter spends some time trying to unpack and make sense of the academic debate, the real significance of the growth in academic analysis of populism for my purposes is that it is trying to describe what seems to be an increasingly important and prominent political phenomenon. Many people are plainly unhappy with the way in which politics is conducted and the apparent inability of politicians, even those who enjoy a democratic mandate and the legitimacy we might expect that to confer, to deal effectively with some of the major problems we seem to confront.[2] Although this book is motivated by the idea that some forms of populism might actually play a constructive role in encouraging policymakers to take climate change seriously, it is important to recognize at the outset that this is *not* usually what populism is about. It is noteworthy and instructive, for example, that the *gilets jaunes* demonstrations were initially sparked by opposition to a proposed eco-tax on polluting, environmentally damaging petrol and diesel.[3] At the very least, this suggests that politically sustainable efforts to win popular support for environmentally friendly policies will need to be part of a more comprehensive debate about public policy and economic opportunity.

Before trying to decide what populism is and whether it has the potential to play a 'progressive' role in politics at either the national or the transnational level, though, it is useful to say something about the principal political alternative to populism in much of the West, at least: democracy. At a time when so many young people in particular appear to have lost confidence in democratic processes and the leaders they produce,[4] it is important to consider whether democratic failure is part of the problem—even if populism is not necessarily the solution.

The Death of Democracy?

Any discussion contemplating the state of contemporary politics needs to make a seemingly obligatory reference to Francis Fukuyama. The one thing that everyone seems to know about Fukuyama is that he famously predicted that a form of free-market, liberal democracy was going to sweep all alternative political and economic systems away and that we had consequently reached 'the end of history'.[5] Educated, individualistic members of the rapidly expanding global middle class would not put up with unrepresentative, repressive autocrats and would demand forms of economic and political liberalism similar to those enjoyed in the US. As we now know, of course, things haven't quite worked out that way and there is a small industry dedicated to pointing this out and emphasizing how wrong Fukuyama was.

However, it is important to remember that one of the reasons Fukuyama attracted such attention was that when he wrote his original article just as the Cold War was ending, plenty of people thought he was right. The world really did seem on the brink of momentous and largely positive change. None of the realists who helped shape America's and the Soviet Union's strategic policies saw the end of the Cold War coming, it should be noted.[6] On the contrary, most of the influential thinkers who shaped strategic policy could imagine nothing other than an enduring balance of power and an endless military and *ideological* standoff with the Soviet Union. Little wonder that so many people were astounded when the Soviet Union disappeared without a shot being fired and some form of capitalism looked like the only economic game in town.

We should also remember that democracy was on something of a roll for a while, too. The late Samuel Huntington was another prominent scholar who attracted more than his fair share of opprobrium, mainly from people unfamiliar with the range and extent of his work. One of Huntington's many important contributions was *The Third Wave*, which detailed the 'global democratic revolution' that seemed to be sweeping all before it from the mid-1970s to the early 1990s.[7] It's not really surprising that so many people thought that there was something historically unprecedented and transformative about the times; I know I did. It did seem for a fleeting historical moment that economic development was all that really mattered, and that an old-fashioned obsession with geopolitics was passé and inappropriate.[8] Significantly, economic development was also seen as the key to underpinning long-run and possibly unstoppable political change.[9] In other words, many analysts thought that rising living standards would help to bring about an equally positive transformation of the political sphere. It was a belief that reflected the European experience of the nineteenth and twentieth centuries.[10]

Even if the claim about the link between economic and political development was accurate, the fact that there appear to be implacable biophysical constraints to an economic model based on unstoppable expansion and the utilization of finite resources raises difficult questions about the sustainability of *political* as well as environmental practices. Likewise, even if there is an unambiguous link between economic development and a move toward political pluralism—and as the case of China demonstrates,[11] this link is far from clear—it looks increasingly difficult to achieve; or it does under the prevailing model of capitalist development, at least.[12] But these sorts of 'structural' constraints are not the only factors that have brought about a 'crisis in democracy' and predictions that it cannot survive.[13] There is also an equally fundamental question about 'agency', highlighted by the fact that across the Western world confidence in democratic processes and representation has been profoundly undermined. A series of economic crises and the failure of elites to deal with them; a perception that the much-invoked concept of the 'level playing field' is tipped against the majority; the sense that national politics is being subverted by external pressures and forces; and the inability of democracies to deal with new challenges such as climate change have all contributed to the current, increasingly widespread loss of confidence in democracy and the rise of alternatives.[14]

Even those who consider that some form of democracy remains both normatively desirable and practically useful concede that it is mutating in the face of powerful new transnational forces. John Keane suggests that

> [t]he old meaning of democracy as "the organization of peaceful competition to hold the reins of power", the periodic election of representatives based on the rule of one person, one vote is dying. It is being superseded by a type of democracy guided by a different and more complex rule: one person, many interests, many votes and chosen by representatives, both at home and abroad.[15]

The appropriate response, Keane argues, is to develop a form of 'monitory democracy' with which to hold elected officials accountable and make their actions transparent. This is especially the case when unfolding environmental crises mean that 'contemporary civil societies are for the first time forced to come to terms self-reflexively with their co-dependence upon, and dwelling within, the biosphere.'[16] A number of other authors have argued that this sort of 'reflexivity' needs to be part of national and international deliberations about how to respond to climate change in particular, although they acknowledge that 'there is a long way to go before that capacity measures up to magnitude of the challenge.'[17] Indeed, the potential for developing some sort of Habermasian ideal speech environment,[18] of a sort that underpins 'deliberative democracy', and its assumption of 'an ideal of "political justification" requiring free public reasoning of equal citizens,'[19] looks formidably difficult.

No matter how well-intentioned and even inspiring some of these ideas may be, the sobering social reality would seem to be that the preconditions in which deliberative democracy can actually occur are challenging, no matter how desirable such outcomes might be in theory.[20] Indeed, one of the dilemmas facing those who would like to encourage a more 'cosmopolitan' and thoughtful engagement with environmental problems,[21] is not simply that such problems are difficult for nonexperts to understand and debate, but that there is an innate distrust—even hostility—on the part of many 'ordinary' people about the nature of expertise. As Frank Fischer points out, there is 'no such thing as a totally free space for deliberation, the world we live in is highly unequal, conflict is as prevalent as consensus, and the mode of argumentation based on the giving of reasons would seem to leave out other important considerations'.[22] Likewise, as Shapiro observed long before the latest wave of populism threatened to wash away the foundations of democracy,

> To the extent that deliberation would be a healthy thing in the formation of public policy, the principal obstacle often is not the lack of will on the part of the people with differing moral convictions to deliberate in ways that can minimize their differences. Rather, the obstacle results from decisions by powerful players who make it their business to shape the terms of public debate through the financial contributions they make available to politicians and political campaigns.[23]

The perception that the political game is rigged is increasingly widespread, and with good reason; and not just at the national level.[24] As we saw in Chap. 3, some voices are listened to much more closely than others when it comes to shaping and implementing policy. It is precisely this apparent inability to affect the policymaking process and the feeling that not all interests are being represented that underpins the surge of interest in populism.

Populism in Theory and Practice

While there is no shortage of voices proclaiming that 'populism is on the march',[25] especially in the West, populism is actually more difficult to define than we might expect or hope given all the attention it has received of late. Although we may think we know it when we see it—Juan and Eva Perón in Argentina, Hugo Chavez in Venezuela, Viktor Orban in Hungary,

Silvio Berlusconi in Italy, perhaps—it has proved surprisingly difficult to develop a definition about which there is universal agreement. It was ever thus in academia, of course, especially in the social sciences. Nevertheless, there is a degree of consensus about the idea that populism reflects and helps to entrench fundamental divisions in *national* societies. While we are arguably yet to see a coordinated and consequential form of transnational populism, Cas Mudde's much cited definition highlights populism's predominantly national dynamics when he suggests that populism is 'an ideology that considers society to be ultimately separated into two homogeneous and antagonistic groups, 'the pure people' versus 'the corrupt elite', and which argues that politics should be an expression of the *volonté générale* (general will) of the people'.[26]

It's not hard to see why the idea of populism seems to capture something distinctive and important about the nature of contemporary politics in many parts of the world, and especially in the US. After all, Donald Trump made a virtue of his alleged status as an outsider and political novice when he ran for office, promising to 'drain the swamp' in Washington, which he claimed had been captured by vested interests and self-serving, unrepresentative elites.[27] The fact that Trump has subsequently filled his administration 'with a mix of the staggeringly wealthy and the staggeringly reactionary'[28] has done little to dampen the enthusiasm of his electoral base, it seems. On the contrary, the widely noted polarization and dysfunctional nature of American politics appears to be getting worse rather than better,[29] which is both a symptom and a cause of the populist resurgence in the US.

The US is, of course, a special case, and one in which many of its inhabitants have seen themselves as occupying a unique historical place. Part of America's self-appointed 'mission' has been the promotion and embodiment of particular virtues, not the least of which are democracy and individual liberty.[30] Whatever one thinks of these values and the efforts of various administrations to promote them at home and abroad, the fact that the Trump administration has made a virtue of putting 'America first' and adopting a more 'transactional' approach to foreign policy is potentially very significant and may actually facilitate the recent rise of populist leaders and authoritarian powers elsewhere, too.[31] The combination of American retrenchment and a diminished commitment to the values that American leaders frequently proclaimed, even if they did not always abide by them,[32] has opened up a more permissive space in which aspiring despots can flourish.[33]

While it is true that different American administrations have supported various populist and/or authoritarian leaders over the years when it suited them to do so,[34] the fact the Trump administration is not even offering rhetorical support for American virtues and the benefits of democracy is contributing to a changed international environment in which values and principles are becoming more instrumental and self-serving.[35] One of the reasons that democracy prospered was because it did so in a supportive geopolitical environment, especially during the Cold War.[36] These days, grand visions of a more just, inclusive and liberal world order have rather gone out of fashion.[37]

Counter-intuitively, perhaps, this is why populism may be the 'ism' for the times, despite the fact that some observers claim it is distinguished by an absence of the sort of coherent ideological perspective that distinguishes liberalism, Marxism and even realism.[38] Mudde and Kaltwasser argue that populism has a 'thin centered ideology' and consequently is best understood as 'a kind of mental map through which individuals analyze and comprehend political reality'.[39] This helps to explain what Taggart describes as populism's 'chameleonic quality', in which it 'always takes on the hue of the environment in which it occurs.'[40] This also suggests why populism need not be exclusively associated with right-wing politics, although historically it generally has been.[41] The protean, contingent and frequently ephemeral nature of populism leads Aslanidis to argue that populism is consequently best thought of as a 'framing device', in which it 'modestly becomes a discourse, invoking the supremacy of popular sovereignty to claim that corrupt elites are defrauding "the People" of their rightful political authority. It becomes an anti-elite discourse in the name of the sovereign People'.[42]

WE, THE PEOPLE

Sovereignty and the role of the people, especially with a capital 'P', are important themes in the study and practice of populism. Indeed, 'the key strength of a populist social movement relies on its capacity to interpret a widespread feeling of anger with the establishment and to convincingly propose that the solution lies in the sovereign people'.[43] At its most extreme, concerns about the inviolability of the nation and the uniqueness of the people can coalesce to form the basis of virulent nativism, even fascism—a possibility most vividly illustrated in the rise of Nazi Germany, of course.[44] Yet even populism's less morbid manifestations are frequently preoccupied with immigration and the protection of national borders. It is no coincidence that Trump has made building a wall on the Mexican border such a priority—or that it remains popular with his 'base', despite its rejection more generally.[45] While it might be tempting to dismiss Trump supporters as bigots and racists, this would miss something important about the underlying drivers of populism's appeal and the reason it has become so prominent in the US and Europe in particular.

In the US, one of the most important factors underpinning Trump's rise and the increased partisanship in domestic politics has been a secular decline in the relative living standards of many Americans.[46] Not only have living standards stagnated and the prospect of achieving the 'American Dream' receded into the distance, but there has also been a dramatic increase in economic inequality as 'the elites' have manipulated the political and economic system to their own advantage.[47] Many formerly well-paid blue collar jobs have disappeared overseas and are unlikely to return, Trump's rhetoric notwithstanding.[48] At the same time, legal and illegal immigrants seem to be placing downward pressure on domestic wages for the skilled and unskilled jobs that do remain, although there is a good deal of highly politicized controversy about this issue.[49] There is often a direct economic cost of such changes for what sociologists used to call the 'labor aristocracy'; but there is also a loss of social standing, dignity and self-worth. None of these issues, it seemed, were taken seriously by the prosperous liberal elite. As Michael Sandel points out, 'the hard reality is that Donald Trump was elected by tapping a wellspring of anxieties, frustrations, and legitimate grievances to which the mainstream parties have no compelling answer'.[50]

The European situation is more complex but concerns about border control and growing levels of illegal migration have been central to the rise of right-wing populist parties from Budapest to Birmingham. Angela Merkel's well-intentioned suggestion that Germany could absorb millions of asylum seekers and refugees from the Middle East and Africa in particular proved to be politically unsustainable.[51] Even Sweden, long the benchmark of progressive politics and generous public policy, has been forced to backtrack on its commitment to assist asylum seekers in the face of a growth in right-wing political activism.[52] Such developments suggest that there are two things that are important to recognize, no matter how unpalatable they may for those people fortunate enough to be members of a global elite with progressive, cosmopolitan values.[53]

First, the division between many of 'the people' and the elites who make policy and—in democracies, at least—purport to represent them is wide and growing. Telling people what they ought to think, or dismissing them as ignoramuses when they don't, is only likely to cement such divisions and increase the appeal of those who claim to be speaking on behalf of the marginalized and neglected.[54] The frequent association of populism with charismatic leadership becomes more understandable in such circumstances. It is one of the claims of populist leaders that they are able to reconcile the tension at the heart of democracy and what Canovan calls its 'pragmatic' and 'redemptive' elements:

> Democracy's redemptive promise is power to the people; we, the people, are to take charge of our lives and to decide our own future. Unfortunately, this promise is in deep and inescapable conflict with democracy viewed in the cold light of pragmatism, and the gap between the two is a fruitful breeding-ground for populist protest.[55]

In other words, despite the distrust many populists feel about contemporary forms of government and their lack of inclusiveness, reliance on technocratic expertise seems like an inescapable feature of governance in complex societies, a possibility that is especially acute and visible in the management of market economies.[56] As Achen and Bartels suggest, 'effective democracy requires an appropriate balance between popular preferences and elite expertise'.[57]

Second, any set of public policies is developed and implemented within specific, historically contingent circumstances. No matter how normatively desirable some ideas and principles may be, if they are misunderstood or simply rejected by large numbers of people—perhaps even a majority—then they are unlikely to be enacted. The views of a small majority of British people, for example, was enough to trigger 'Brexit', and profoundly change that country's domestic and international position, in ways that they may come to regret—or they will according to the technocrats at the Bank of England, at least.[58] Inglehart and Norris argue that such an outcome was not as surprising as many elites found it, but reflects the fact that 'the classic economic Left-Right cleavage in party competition is overlaid today by a new cultural cleavage dividing populists from cosmopolitan liberalism.'[59] Consequently, they conclude that 'cultural values' in combination with social and demographic change account for the rise in support for populist political parties.

At one level, there is no doubt that nations are distinguished by different cultures and values and that these have proved surprisingly durable. Expectations about the disappearance of 'premodern' traditions and values have often proved to be misplaced wishful thinking. Indeed, there has been a virulent backlash against 'Western modernity', most visible in the complex radicalization of Islam as its adherents around the world struggle to come to terms with values that challenge many of its essential beliefs and traditions.[60] And yet it is not only more 'traditional' societies that are being transformed by many of the processes that are subsumed under the convenient shorthand of 'globalization'.[61] On the contrary, Western societies are much less cohesive and culturally homogenous than they once were, the nostalgic appeals to populist national identities notwithstanding. One of the most divisive and disempowering features of the contemporary political scene in many of the 'advanced' industrial democracies has been the decline of traditional forms of social interaction and organization. This is nowhere more evident than in the declining appeal and membership of traditional political parties.[62]

THE INDIVIDUAL VERSUS THE COLLECTIVE

One of the more striking features of contemporary social and political life in much of the Western world has been the rise of identity politics, and a growing sense of individual rather than collective identity. In some ways, there is nothing new about this possibility, as Christopher Lasch pointed out some 40 years ago.[63] What *is* new and distinctive, though, is the impact this is having on some more traditional forms of political representation. One result of the loss of manufacturing sector jobs in particular is not simply that they contribute to the declining economic prospects of the working class, but that this process accelerates the erosion of the working class as a concept and source of identity, as well as the basis for political mobilization. The way many people think about themselves and their relationship to each other and to the political system that notionally represents their interests has changed as a consequence, too.[64]

The casualization and fragmentation of the workforce have both contributed to the decreased importance of the trade union movement and the sense of collective social identity such organizations engendered.[65] Some prominent social scientists have long bemoaned the decline of 'social capital' as part of the explanation for a range of problems from a diminished sense of social solidarity, a reduction in the ease of doing business, or of establishing relationships based on trust and reciprocity.[66] The much-maligned Professor Fukuyama has been at the forefront of analyses of these developments, too, not only usefully distinguishing between high and low trust societies,[67] but helping to explain why identity politics has been one of the key manifestations of widespread social transformation.[68]

It is not simply the fact that many people—especially the young upon whom so much depends—now find the old categories of class and nationality less compelling and meaningful than their forbears that is potentially problematic when it comes to large scale collective action. Many of the new forms of identity and social mobilization revolve around increasingly narrow questions of ethnicity and gender, which have limited appeal to many outside of the urban liberal heartlands of North America in particular. On the contrary, a visceral antipathy toward new social issues is entrenching the growing divisions of wealth, opportunity and outlook that exist between the educated, cosmopolitan elites who reside in places like New York and London, and their less skilled, less mobile counterparts in the agricultural hinterlands and the deindustrializing rust belts of former manufacturing powerhouses.[69]

According to Fukuyama, there are a number of features of the new politics of identity that potentially militate against effective social mobilization and collective action. On the one hand, the growth of identity politics on the 'Left' has encouraged a similar process on the 'Right'. Whether the relationship is entirely straightforward is debatable, but it is not controversial to suggest that nationalism and concerns about the erosion of established identities and social standing have been central to the rise of populist politics in the US and Europe.[70] Whatever the merits of normatively admirable ideas such as multiculturalism for example, it presents a potentially major challenge to many members of the formerly privileged white working class in particular. As Eric Kauffmann points out, in parts of the UK with high levels of immigration, 'it is difficult to see how members of the ethnic majority were supposed to fit into the notion of multiculturalism except as apologists for past misdeeds'.[71]

Again, it is not necessary to approve of or even sympathize with the rather xenophobic reactions of those who feel—rightly or wrongly—dispossessed and discomfited by major structural, social and political changes over which they have little control. We do, however, have to recognize that it matters and represents a potentially major obstacle to collective action of any sort, but especially for 'progressive' causes. Winning the support of those who feel marginalized rather than empowered by globalization and the transformation of the old order is a formidable but possibly inescapable challenge. Yet, not only are political activists on the contemporary Left generally not sympathetic to trade unionism and the privileged position of the old working class, Fukuyama argues, but identity politics may also undermine the basis of many of the achievements of 'the West', the Enlightenment and the general—increasingly problematic and vilified—process of modernization:

> [T]he left's identity politics poses a threat to free speech and to the kind of rational discourse needed to sustain a democracy. … The focus on lived experience by identity groups prioritizes the emotional world of the inner self over the rational examination of issues in the outside world and privileges sincerely held opinions over a process of reasoned deliberation that may force one to abandon prior opinions.[72]

By this stage, the reader may be forgiven for wondering how the foregoing discussion impacts on my central claim that populism might offer a way of creating sufficient political pressure to compel political leaders to act on climate change while they still can. Unfortunately, the pessimists—whether they are taking about climate change or the possibilities for collective political and social action to address it—do seem to have some of the most compelling arguments. Whatever we may think about these arguments or the people that make them, they need to be taken seriously if 'we' are to make a difference and develop more inclusive strategies for action. Consequently, I'm afraid, the entire book is written more in hope than in expectation. But before actually considering what the prospects for good and bad populism might be, it is important to make a few brief remarks about the potential for positive, even progressive populism.

The Way We Were (and Could Be)?

The possibility that populism might not be an exclusively right-wing or reactionary phenomenon has been taken up by social theorists working in a broadly Marxist tradition. At the very least, this is interesting in theoretical terms and suggests that it is important not to write off the potentially transformative impact of social movements and leaders on the broadly conceived left wing of politics. Whether they fit some of the standard definitions of populism or not, the popularity of figures such as Bernie Sanders in the US and Jeremy Corbyn in the UK demonstrates that political mobilization around 'radical' figures is possible. The remarkable impact of the charismatic, telegenic and *young* Alexandria Ocasio-Cortez, a newly elected member of the House of Representatives, demonstrates the potential that a compelling reformist discourse can actually have in even a dysfunctional and polarized political system.[73]

Unfortunately, the work of some scholars on the Left who are supposedly interested in deconstructing and accounting for the operation of dominant discourses is often remarkably impenetrable. Nevertheless, Ernesto Laclau argued that populism was best thought of as a 'political logic', and not necessarily exclusively associated with either Left or Right, but with 'the establishment of an area of variations within which a plurality of phenomena could be inscribed'.[74] Chantal Mouffe, Laclau's frequent collaborator,[75] endorsed the importance of discourse in thinking about populism, and suggested that the objective of a Left populism ought to be the 'construction of a collective will, a "people" apt to bring about a new hegemonic formation that will reestablish the articulation between liberalism and democracy that has been disavowed by neoliberalism, putting democratic values in the leading role'.[76]

Two points are worth making about this, in addition to the fact that prominent intellectuals on the Left see populism as a potential vehicle for possible social mobilization. First, although 'neoliberalism' is frequently seen as the root cause of many of the world's environmental problems,[77] the reality is more complex: many parts of the world—such as China, for an important example—are decidedly *not* neoliberal, and the causes of many of our collective ecological problems would seem to have more to do with planetary boundaries, the carrying capacity of the Earth and overconsumption.[78] The second issue that merits emphasis is about the construction of dominant discourses themselves, and here the critical theorists do seem to have a point. Not only are consumerist lifestyles and unsustainable consumption patterns relentlessly promoted within capitalist economies, but the ability of vested interests to control what is disseminated by mainstream media powerfully shapes public attitudes toward important policy issues.[79] As we have seen, this is not simply a problem that affects the West either: China reminds us that social media can be controlled and used for political repression rather than liberalization.[80]

And yet we also need to recognize that new forms of social media *have* played a crucial role in facilitating social mobilization and collective action, as we shall see in more detail in Chap. 5. The short-lived 'Arab Spring' is a salutary and illustrative lesson of both the potential and the difficulties facing would-be revolutionaries or agents of social change everywhere, but especially in parts of the world where authoritarian repression is the familiar order of the day.[81] The point to emphasize is that establishing and sustaining an independent, effective civil society is difficult enough at the national level, let alone globally. For all the hopes that were invested in the potentially transformative nature of 'global civil society', especially in relation to the environment,[82] thus far it has had limited impact, especially in those parts of the world with little history of widespread political emancipation.[83]

SOCIAL MOVEMENTS AND GLOBAL CIVIL SOCIETY

The relationship between the national and the transnational is, as we saw in Chap. 3, one of the most important conceptual and practical issues in the current international order. The challenge for environmental activists in particular has been to develop modes of organization and activism that transcend national borders and ultimately combine the efforts of citizens around the world. Given Karl Polanyi's influential thesis about the potentially transformative relationship that exists between human beings and their material conditions of existence in market-ordered societies, the fact that this hasn't happened is, perhaps, more surprising than it may seem.

Polanyi famously claimed that the creation and imposition of an unregulated *laissez faire* economy would create a 'double movement' in which people would demand 'social protection' from the 'deleterious impact of the market'. With remarkable foresight, given that he was writing at the height of the World War II when concerns about the environment were decidedly not on the agenda, such demands would, Polanyi claimed, revolve around the 'conservation of man and nature'.[84] The reason this has not occurred in quite the way that Polanyi expected is, according to a later generation of critical scholars, because the sovereignty of Polanyi's time, which was associated primarily with states, has now become 'imperial'. In this reading, the rule-setting capacity of states has been undermined as its 'constitutional functions have been displaced to another level',[85] leaving nongovernment organizations (NGOs) to play a key role in applying pressure from below—as long as they are not co-opted or actually government-organized NGOs (GONGO), of course.[86]

The reason so many people were optimistic about the role of non-state actors is that they were thought to 'pluralize power and to problematize violence.'[87] This is also why many observers were optimistic about both the pacifying effect of democracy generally and about the progressive regulatory potential of 'cosmopolitan democracy' in particular.[88] As we now know, thus far at least, things haven't quite worked out that way either; this looks like another argument in which the realists have the more compelling evidence. And yet there are still many people making the case that the 'group of people making a real impact in discovering, formulating, and implementing solutions to global problems has expanded dramatically.'[89] Significantly, what Keck and Sikkink describe as 'advocacy networks' are motivated primarily by values rather than material concerns, and in some issue areas, have demonstrated a capacity to 'instigate changes in the institutionalized and principled basis of international interactions.'[90]

While some might quibble about how extensive and durable the changes they detail have proved to be, they do have a point: there already is some compelling evidence of not just the potential for transnational social mobilization and organization, but also its tangible realization. The fact that this has happened at all is perhaps the crucial point, and helps to explain the more widespread, albeit less coherent growth of social movements, which are taken to be another optimism-inducing manifestation of the contemporary era. Unfortunately, and despite the high profile and public attention some social movements have received, it seems the millennial generation's disenchantment with democracy and politics in general also extends to their participation in social movements as well.[91]

Charles Tilly, who did more than anyone to help define social movements (and much else), argued that they had three principal components: campaigns of collective claims on target authorities; an array of claim-making performances, including special purpose associations, public meetings, media statements and demonstrations; and public representations of the cause's worthiness, unity, numbers and commitment.[92] This is a potentially demanding agenda in a national context. To develop transnational social movements with the potential to address similarly transnational challenges is 'not easy', as Sidney Tarrow laconically points out: 'sustaining collective action across borders on the part of people who seldom see one another and who lack embedded relations of trust is difficult.'[93] Recent history suggests it is not impossible, though, even if translating energy and activism into actual policy remains a problem.

However, there is another possibility that is not always given the attention it merits by scholars of social movements and transnational activism: they are not necessarily progressive.[94] Populist movements that transcend national borders or which encourage emulation elsewhere are clearly one possible example of this, but so too is radical Islam, which has become an influential social force for which people are prepared to give their lives. Motivation and enthusiasm for a cause doesn't get much more sincere than this. The terrorist networks associated with this movement are also important examples of the way that the technologies and opportunities of the increasingly interconnected world can also be used for profoundly anti-modern and destructive purposes.[95]

One other aspect of social movements needs to be acknowledged: some issue areas are much more susceptible to normative and policy change than others. The widely praised and much-discussed campaign to rid the world of landmines was highly successful and effective not least because the opposition to banning their use was bereft of credible arguments.[96] The very existence of something resembling global civil society effectively changed the nature of domestic debates and helped this movement realize its goals. But what of other issue areas, where the forces blocking reform are more entrenched, and where the required changes amount to nothing less than the transformation of practical and conceptual place of the dominant economic system? As Naomi Klein observes, 'given that the challenge for the climate movement hinges on pulling off a profound and radical *economic* transformation, it must be noted that for these new movements, the legal and cultural battles were always more successful than the economic ones'.[97]

Again, it might seem that the structural and agential obstacles ranged against environmental activists and the development of any form of transnational collective action are formidable. Indeed, they are. It is for this reason that any global movement has first to be grounded in local, national communities, before it can turn its collective attention to the wider problems of global society. Even advocates of cosmopolitan attitudes and a transcendence of national sovereignty, such as Andrew Linklater, recognize that 'the state remains the principal site on which efforts to monopolize the control of significant resources and opportunities and struggles to create less exclusionary communities is worked out.'[98] Put differently, unless the domestic political environment is conducive to progressive reform and actually addressing ecological problems, our chances of bringing about significant change at the global level look bleak. We have to start making the necessary changes somewhere, and it is best to act where we have some chance of making a difference. If we can't fix our problems and exercise our own sense of agency and responsibility, why would we expect to influence anyone else's?

It is for this reason that thinking globally but *acting* locally still seems like an idea with merit.[99] The chances of doing so in democracies—for all their widely noted shortcomings—are at least good in principle. Political action and engagement may even offer a way of realizing the transformative potential of populism. In this context, Albert Dzur and Carolyn Hendriks make an interesting and potentially useful distinction between what they describe as 'thin' and 'thick' forms of populism. Thin and thick populism are distinguished by their different attitudes to similar issues: Thin populism sees government as intrusive; experts as untrustworthy; is preoccupied with mobilizing to trigger policy and leadership change; and has short-term time horizons. Thick populism, by contrast, sees extant governments as potentially useful; is open to collaboration with experts; wants to reform and improve institutions rather than overthrow them; and has a longer-term time horizon. Equally importantly, thick populism is less obsessed about racial homogeneity and individualism, and willing to work collaboratively, even in multicultural settings.

Thin populism is, therefore, closely related to the sort of 'thin centered ideology' discussed earlier and refers to the mobilization of popular support around particular issues, often using a discourse centered on race, the nation or the will of 'the people' as its justification and rallying cry. Thick populism, by contrast, 'promotes popular participation through the organizing strategies of citizens' movements working to circumvent, challenge, and reform political institutions.'[100] In this perspective, democracy and existing political institutions actually have the potential to be part of the solution, rather than the problem, as so many participants in thin versions of populism believe. Indeed, Dzur and Hendricks argue that 'contemporary populist efforts might, with the right forms of organization, become institutionally redemptive, rather than destructive, leading to a sense of ownership, rather than further distrust and alienation.'[101]

Conclusion

Skeptical readers may be forgiven a little eye-rolling at this point. After all, most of this chapter has been taken up with outlining some of the problems that confront traditional democracy, social movements and people interested in tackling unprecedented transnational problems. In this context, thick populism looks suspiciously like a pragmatic, policy-oriented version of deliberative democracy. For all its many admirable principles and intentions, the conditions in which any sort of deliberative democracy could actually occur—even in a less turbulent world than our own—look forbiddingly difficult and unlikely. As Arias-Maldonado pithily observes, 'the politics of nature cannot escape the nature of politics', with the implication being that 'the belief that citizens in a deliberative context will spontaneously acquire ecological enlightenment, and will push for greener decisions, relies too much on an optimistic, naive view of human nature, so frequently found in utopian political movements.'[102]

Perhaps so. To be sure, I share many of the doubts of those who fear the worst and question democracy's capacity to find a solution or a mechanism for action.[103] The fact that much of the world appears to be moving away from, rather than toward, democracy is a major problem, even in a less constrained time frame than we currently confront. Yet as unlikely as many of the potential 'solutions' may look, we must hope that some of them contain the seeds of effective action. Without effective action, some of the more apocalyptic visions, in which a 'matrix of parasitic interests has begun to shape adaptation as the militarized management of civilization's violent disintegration',[104] don't look quite as paranoid or delusional as we might hope. As we saw in Chap. 3, security agencies around the world are actively contemplating the potentially catastrophic consequences of climate change and working out how their national interests and communities will be best served. Claims that we are all in this together, no matter how true they may be, are unlikely to cut much ice with policymakers and strategic elites who subscribe to a zero-sum, state-centric view of the world.

And yet, as we shall see in Chap. 5, there is evidence of something like thick populism in action. It is at least possible to make a case for cautious and qualified optimism. This may involve what the Italian Marxist theorist Antonio Gramsci described as a triumph of the 'optimism of the will' over the 'pessimism of the intellect'.[105] Indeed, as Mrs. Thatcher was wont to remark, albeit in rather different circumstances, 'there is no alternative' when it comes to taking climate change seriously and trying to do something about it.

Notes

1. Moffitt, B. (2016) *The Global Rise of Populism: Performance, Political Style, and Representation* (Stanford: Stanford University Press).
2. Hay, C. (2007) *Why We Hate Politics* (Cambridge: Polity).
3. Bremner, C. (2018) 'Macron reverses on fuel tax increases'. *The Australian* December 4.

4. Bremmer, I. (2018) 'Is democracy essential? Millennials increasingly aren't sure—and that should concern us all', *Think*, NBCNews, February 13. https://www.nbcnews.com/think/opinion/democracy-essential-millennials-increasingly-aren-t-sure-should-concern-us-ncna847476
5. Fukuyama, F. (1992) *The End of History and the Last Man* (New York: Free Press).
6. Lebow, R.N. (1994) 'The long peace, the end of the cold war, and the failure of realism', *International Organization* 48(2): 249–277.
7. Huntington, S. (1991) *The Third Wave: Democratization in the Late Twentieth Century* (Norman: University of Oklahoma Press).
8. Luttwak, E. (1990) 'From geopolitics to geo-economics', *The National Interest* (Summer): 17–23.
9. Przeworski, A., Alvarez, M.E., Cheibub, J.A. and Limongi, F. (2000) *Democracy and Development: Political Institutions and Well-Being in the World, 1950–1990* (Cambridge: Cambridge University Press).
10. Rueschemeyer, D., Stephens, E.H. and Stephens, J.D. (1992) *Capitalist Development and Democracy* (Cambridge: Polity Press); Moore, B. (1966 [1987]) *Social Origins of Dictatorship and Democracy* (Penguin).
11. Tsai, K.S. (2007) *Capitalism Without Democracy: The Private Sector in Contemporary China* (Ithaca: Cornell University Press); Wright, T. (2010) *Accepting Authoritarianism: State-Society Relations in China's Reform Era* (Stanford: Stanford University Press).
12. Hsueh, R. (2016) 'State capitalism, Chinese-style: Strategic value of sectors, sectoral characteristics, and globalization', *Governance* 29(1): 85–102.
13. Runciman, D. (2018) *How Democracy Ends* (London: Profile Books).
14. Mounk, Y. and Foa, R.S. (2018) 'The End of the democratic century: Autocracy's global ascendance', *Foreign Affairs* 97(3): 29–36.
15. Keane, J. (2018) *Power and Humility: The Future of Monitory Democracy* (Cambridge: Cambridge University Press), p. 14.
16. Keane, ibid., p. 232.
17. Dryzek, J.S. and Pickering, J. (2019) *The Politics of the Anthropocene* (Oxford: Oxford University Press), p. 43.
18. Jürgen Habermas developed the idea of an 'ideal speech situation', governed by socially constituted rules that allowed reasoned, open-ended debate between informed participants. See, Habermas, J. (1985) *The Theory of Communicative Action: Reason and the Rationalization of Society* (Boston: Beacon Press).
19. Bohman, J. (1998) 'Survey article: The coming of the age of deliberative democracy', *Journal of Political Philosophy* 6(4), p. 402.
20. Eckersley, R. (2000) 'Deliberative democracy, ecological representation and risk', in, M. Saward *Democratic innovation: Deliberation, representa-*

tion and association (London: Routledge): 117–132; Smith, G. (2003) *Deliberative Democracy and the Environment* (London: Routledge).
21. Dobson, A. (2006) 'Thick cosmopolitanism', *Political Studies* 54(1): 165–184.
22. Fischer, F. (2009) *Democracy and Expertise: Reorienting Policy Inquiry* (Oxford: Oxford University Press), p. 82.
23. Shapiro, Ian (2003) *The State of Democratic Theory* (Princeton: Princeton University Press), p. 30.
24. Colgan, J. D., & Keohane, R. O. (2017) 'The liberal order is rigged.', *Foreign Affairs*, 96(3), 36–44.
25. Zakaria, F. (2016) 'Populism on the march: Why the West is in trouble', *Foreign Affairs*, 95 (6), pp. 9–15.
26. Mudde, C. (2004) 'The populist zeitgeist', *Government and Opposition*, 39(4), 542–563.
27. Beeson, M. (2016) 'Trumpian deswampification', *The Conversation*, December 8.
28. Pierson, P. (2017) 'American hybrid: Donald Trump and the strange merger of populism and plutocracy', *British Journal of Sociology* 68, p. S107.
29. Luce, E. (2018) 'America's political journey into tribalism'. *Financial Times* January 24; Mann, T.E. and Ornstein, N.J. (2012) *It's Even Worse Than It Looks: How the American Constitutional System Collided with the New Politics of Extremism* (New York: Basic Books).
30. McDougall, W.A. (1997) *Promised Land, Crusader State: The American Encounter with the World Since 1776* (Boston: Mariner Books); Smith, T. (1994) *America's Mission: The United States and the Worldwide Struggle for Democracy in the Twentieth Century* (Princeton: Princeton University Press).
31. Beeson, M. (forthcoming) 'Donald Trump and post-Pivot Asia: The implications of a 'transactional' approach to foreign policy', *Asian Studies Review*.
32. Johnson, C. (2004) *The Sorrows of Empire: Militarism, Secrecy, and the End of the Republic* (New York: Metropolitan Books).
33. Gat, A. (2007) 'The return of authoritarian great powers', *Foreign Affairs* 86(4): 59–69; Rachman, G. (2011) *Zero-Sum Future: American Power in an Age of Anxiety* (New York: Simon & Schuster).
34. Bacevich, A.J. (2002) *American Empire: The Realities and Consequences of US Diplomacy* (Cambridge, Mass: Harvard University Press).
35. Eatwell, Roger and Goodwin, Matthew (2018) *National Populism: The Revolt Against Liberal Democracy* (London: Pelican), p. 109.
36. Narizny, K. (2012) 'Anglo-American primacy and the global spread of democracy: An international genealogy', *World Politics* 64(02): 341–373.

37. Haass, R. (2018) 'How a world order ends and what comes in Its wake', *Foreign Affairs* (January/February).
38. Realists might like to think that their perspective is nonideological and simply a factual description of reality, but this would be to neglect the way in which knowledge is socially constructed and a reflection of the power and ideas of society's most powerful actors. See Carstensen, M.B. (2011) 'Paradigm man vs. the bricoleur: bricolage as an alternative vision of agency in ideational change', *European Political Science Review* 3(1): 147–167; Legro, J.W. (2005) *Rethinking the World: Great Power Strategies and International Order* (Ithaca: Cornell University Press).
39. Mudde, C. and Kaltwasser, C.R. (2017) *Populism: A Very Short Introduction* (Oxford: Oxford University Press), 6.
40. Taggart, P. (2000) *Populism* (Buckingham: Open University Press), p. 4.
41. Bonikowski, B. (2017) 'Ethno-nationalist populism and the mobilization of collective resentment'. *British Journal of Sociology*, 68, S181–S213.
42. Aslanidis, P. (2015). Is populism an ideology? A refutation and a new perspective. *Political Studies*, 64, p. 96.
43. Mudde, and Kaltwasser, *op cit*, pp. 47/48.
44. Berman, S. (2016) 'Populism is not Fascism: But it could be a harbinger', *Foreign Affairs* October 16.
45. Cohn, N. (2019) 'The wall is not popular. (And neither is Trump.)', *New York Times*, January 12.
46. DeSilver, D. (2018) 'For most U.S. workers, real wages have barely budged in decades', *Facttank: Pew Research Centre*, August 7.
47. Pierson, *op cit*, p. S106.
48. Long, H. (2016) 'U.S. has lost 5 million manufacturing jobs since 2000', *CNN Business*, March 29.
49. *The Economist* (2016) 'Wage war', August 25.
50. Sandel, M. J. (2018). Populism, liberalism, and democracy. *Philosophy and Social Criticism*, 44(4), p. 354.
51. Sharkov, D. (2018) 'Merkel backs down, agreeing to stem flow of refugees into Germany, *Newsweek*, July 3.
52. Eatwell, Roger and Goodwin, Matthew (2018) *National Populism: The Revolt Against Liberal Democracy* (London: Pelican), pp. 36–37; Betz, H.-G. (1994) *Radical Right-Wing Populism in Western Europe* (London: Macmillan).
53. Ulaş, L. (2016) 'Cosmopolitanism, self-interest and world government', *Political Studies* 64(1S): 105–120.
54. Eatwell & Goodwin, *op cit*, p. 109.
55. Canovan, M. (1999). 'Trust the people! Populism and the two faces of democracy'. *Political Studies*, 47(1), p. 11.
56. Snyder, J. (2017) 'The modernization trap', *Journal of Democracy* 28(2): 77–91.

57. Achen, C.H. and Bartels, L.M. (2017) *Democracy for Realists: Why Elections Do Not Produce Responsive Government* (Princeton: Princeton University Press), p. 303.
58. Partington, R. (2018) 'Bank of England says no-deal Brexit would be worse than 2008 crisis', *The Guardian*, November 29.
59. Inglehart, R., & Norris, P. (2016) 'Trump, Brexit, and the Rise of Populism: Economic Have-Nots and Cultural Backlash (No. 16–026). *Faculty Research Working Paper Series.*
60. Roy, O. (2006) *Globalized Islam: The Search for a New Ummah* (New York: Columbia University Press); Barber, B.R. (2001) *Jihad* vs. *McWorld* (New York: Ballantine Books).
61. There is a vast literature on this topic, but its impact on social values is usefully discussed in Appadurai, A. (1996) *Modernity at Large: Cultural Dimensions of Globalization* (Minneapolis: University of Minnesota Press); Meyer, J.W., Boli, J., Thomas, G.M. and Ramirez, F.O. (1997) 'World society and the nation-state', *American Journal of Sociology* 103(1): 144–181; Castells, M. (1997) *Power of Identity: The Information Age: Economy, Society, and Culture* (Oxford: Blackwell Publishers).
62. Dalton, R.J. and Wattenberg, M.P. (2002) *Parties Without Partisans: Political Change in Advanced Industrial Democracies* (Oxford: Oxford University Press).
63. Lasch, C. ([1979] 2018) *The Culture of Narcissism: American Life in an Age of Diminishing Expectations* (New York: WW Norton).
64. Aronowitz, S. (2014) *The Politics of Identity: Class, Culture, Social Movements* (London: Routledge).
65. Kelly, J. (2015) 'Trade union membership and power in comparative perspective', *The Economic and Labour Relations Review* 26(4): 526–544; Siegmann, K.A. and Schiphorst, F. (2016) 'Understanding the globalizing precariat: From informal sector to precarious work', *Progress in Development Studies* 16(2): 111–123.
66. Putman, R.D. (1992) *Making Democracy Work: Civic Traditions in Modern Italy* (Princeton, NJ: Princeton University Press).
67. Fukuyama, F. (1995) *Trust: The Social Virtues and the Creation of Prosperity* (London: Hamish Hamilton).
68. Fukuyama, F. (2018) *Identity: Contemporary Identity Politics and the Struggle for Recognition* (London: Profile).
69. Uberti, D. (2017) 'A divided empire: what the urban-rural split means for the future of America', *The Guardian*, May 11.
70. Norris, P. and Inglehart, R. (2019) *Cultural Backlash and the Rise of Populism: Trump, Brexit, and Authoritarian Populism* (Cambridge: Cambridge University Press).
71. Kaufmann, E. (2018) *Whiteshift: Populism, Immigration and the Future of White Majorities* (London: Allen Lane), p. 155.

72. Fukuyama, F. (2018) 'Against identity politics', *Foreign Affairs* September/October, p. 7.
73. Crooks, E. and Kiran, S. (2019) 'Democrats unveil proposal for "Green New Deal"', *Financial Times*, February 8.
74. Laclau, E. (2005) *On Populist Reason* (London: Verso), p. 175.
75. See, for example, Laclau, E. and Mouffe, C. (1985) *Hegemony and Socialist Strategy* (London: Verso).
76. Mouffe, C. (2018) *For a Left Populism* (London: Verso Books), p. 45.
77. Mouffe argues that 'it is simply impossible to envisage a project of radicalization of democracy in which the "ecological question" is not the center of the agenda'. Mouffe, ibid., p. 61.
78. O'Neill, D.W., Fanning, A.L., Lamb, W.F. and Steinberger, J.K. (2018) 'A good life for all within planetary boundaries', *Nature Sustainability* 1(2): 88–95.
79. McChesney, R.D. (2004) *The Problem of the Media: US Communication Politics in the Twenty-first Century* (New York: NYU Press); Vinton, K. (2016) 'These 15 billionaires own America's news media companies', *Forbes*, June 1.
80. *The Economist* (2018) 'Apartheid with Chinese characteristics: China has turned Xinjiang into a police state like no other', May 31st.
81. Davis, M. (2011) 'Spring confronts Winter', *New Left Review* 72(Nov/Dec): 5–15.
82. Dryzek, J.S., Downes, D., Hunold, C., Schlosberg, D. and Hernes, H.-K. (2003) *Green States and Social Movements: Environmentalism in the United States, United Kingdom, Germany, and Norway* (OUP Oxford). Wapner, P. (1995) 'Politics beyond the state: Environmental activism and world civic politics', *World Politics* 47, 311–340.
83. Teets, J.C. (2014) *Civil Society Under Authoritarianism: The China Model* (Cambridge University Press); Rodan, G. and Jayasuriya, K. (2009) 'Capitalist development, regime transitions and new forms of authoritarianism in Asia', *The Pacific Review* 22(1): 23–47; Mounk, Y. and Foa, R.S. (2018) 'The End of the democratic century: Autocracy's global ascendance', *Foreign Affairs* 97(3): 29–36.
84. Polanyi, K. (1957) *The Great Transformation: The Political and Economic Origins of Our Time* (Boston: Beacon Press), p. 132.
85. Hardt, M. and Negri, A. (2000) *Empire* (Cambridge: Harvard University Press), p. 309.
86. Naim, M. (2007) 'Missing links: What is a GONGO?' *Foreign policy* 160: 95–96.
87. Keane, J. (2003) *Global Civil Society?* (Cambridge: Cambridge University Press), p. 8.
88. Russett, B.M. (1995) *Grasping the Democratic Peace: Principles for a post-Cold War World* (Princeton: Princeton University Press); Held, D. (1995) *Democracy and the Global Order* (Cambridge: Polity Press).

89. Slaughter, A.-M. (2017) *The Chessboard and the Web: Strategies of Connection in a Networked World* (New Haven: Yale University Press), p. 20. Also see Florini, A. (ed., 2000) *The Third Force: The Rise of Transnational Civil Society* (Washington: Carnegie Endowment).
90. Keck, M.E. and Sikkink, K. (1999) *Activists Beyond Borders: Advocacy Networks in International Politics* (Ithaca: Cornell University Press), p. 2.
91. Foa, R.S. and Mounk, Y. (2016) 'The democratic disconnect ", *Journal of Democracy* 27(3), p. 11.
92. Tilly, C. (2015) *Social Movements 1768–2012*, 3rd ed. (London: Routledge), p. 8.
93. Tarrow, S. (2005) *The New Transnational Activism* (Cambridge: Cambridge University Press), p. 7.
94. Munck, R. (2006) *Globalization and Contestation: The New Great Counter-Movement* (London: Routledge), pp. 29–34.
95. Sageman, M. (2011) *Leaderless Jihad: Terror Networks in the Twenty-first Century* (Philadelphia: University of Pennsylvania Press).
96. Cameron, M.A., Lawson, R.J. and Tomlin, B.W. (1998) *To Walk Without Fear: The Global Movement to Ban Landmines* (Toronto: Oxford University Press).
97. Klein, N. (2014) *This Changes Everything: Capitalism* vs. *the Climate* (London: Penguin), p. 453. [Emphasis in original].
98. Linklater, A. (1998) *The Transformation of Political Community* (Polity Press), p. 157.
99. This thesis was most influentially developed in Schumacher, E.F. (1973) *Small is Beautiful: A Study of Economics as if People Mattered* (London: Vintage Boos).
100. Dzur, A.W. and Hendriks, C.M. (2018) 'Thick populism: democracy-enhancing popular participation', *Policy Studies* 39(3), p. 335.
101. Dzur and Hendriks, ibid., p. 338.
102. Arias-Maldonado, M. (2007) 'An imaginary solution? The green defence of deliberative democracy', *Environmental Values* 16(2), p. 248.
103. Shearman, D. and Smith, J.W. (2007) *Climate Change Challenge and the Failure of Democracy* (Westport: Praeger).
104. Parenti, C. (2011) *Tropic of Chaos: Climate Change and the New Geography of Violence* (New York: Nation Books), p. 14.
105. This quote comes from one of Gramsci's prison letters, but has also been attributed to Romain Rolland. Whoever the author may have been, it captures an important point, which may have given Gramsci some comfort: he spent 11 years in prison, which destroyed his health, before his death at age 46.

CHAPTER 5

Populism in Practice

It is all too easy to point to examples of 'bad' populism. Historically, there has been no shortage of authoritarians, megalomaniacs and egomaniacs claiming to represent the people and to be carrying out their will. Even more deflatingly, perhaps, there is no shortage of people who share the beliefs of populist leaders and are willing to offer their support—at least for a while. No amount of evidence about the folly of encouraging wannabe despots is likely to have a prophylactic impact on a people who have worked themselves up into a righteous frenzy about their own significance, the duplicity of elites, the inferiority of outsiders or any of the other nativist, nationalist, nonsense peddled by chauvinists and charlatans. Nor is this sort of populist upsurge an exclusively right-wing phenomenon either. On the contrary, left-wing leaders are equally prone to delusions of grandeur about their place in the historical scheme of things. When you have the adulation and the deaths of millions on your CV, as Mao and Stalin did, perhaps you can be forgiven for taking yourself seriously.[1] Perhaps their admirers couldn't see the bodies piling up; perhaps they were too frightened to speak. Now that we do know, though, you would think we would avoid creating similar circumstances that allow autocrats to flourish. Yet the one thing we learn from history, as they say, is that we don't learn from history.

No doubt, some will quibble about describing Stalin and Mao as populists,[2] but it's important to remember that both of them enjoyed widespread popular support; remarkably enough they still do.[3] Likewise, there's a debate to be had about whether Xi Jinping should be classed as a populist leader[4]; but he is doing everything he can to make himself popular with the masses, not least by embarking on a campaign to make China great again, to borrow a phrase from the world's other superpower.[5] Although not everyone thinks that Donald Trump should be described as a populist either,[6] his relentlessly nationalist agenda and claim to represent 'real' Americans not only ticks some definitional boxes, but also has potentially ominous consequences for the US and the wider world.[7] What unites some of these figures is their assumption that they are in some way the expression of an underlying, hitherto inadequately expressed popular will, and that 'opponents are not just people with different priorities and values, they are evil! Consequently, compromise is impossible, as it "corrupts" the purity'.[8]

This doesn't look like an especially good time to be trying to make an argument that some forms of populism might actually be useful—even necessary—for creating the momentum to address complex problems, let alone the most difficult challenge we have ever faced as a species. But given that nothing else has worked adequately thus far, and that political mobilization and participation are undoubtedly necessary for trying to bring about any sort of major social change, it may be worth a try. Before considering what the 'good' version of populism might look like, however, in the interests of balance and some semblance of academic credibility, it's important to acknowledge what can go wrong, too.

(Mostly) Bad Populism

Arguments and analogies that invoke the name of Adolf Hitler generally aren't taken terribly seriously, so I'll keep my remarks about *Der Fuhrer* fairly brief. The fact that a country that produced the likes of Kant, Hegel, Schopenhauer, Leibniz, Marx, et al., and which had a long history of critical thinking, could go collectively mad still takes some explaining, and provides a very sobering lesson about the possible perils of populism.[9] We forget or minimize such lessons at our peril: if it could happen in Germany, it could happen anywhere. Fortunately, most populist leaders seem to end up running countries that are not as well organized as Germany and tend to do much less damage as a consequence; or they do to the outside world at least. As Mao Zedong reminds us, however, the capacity of uncontrolled megalomaniacs to inflict catastrophic suffering on their own people should never be underestimated.[10]

While the actions of the likes of Mao, Hitler and Stalin have been widely chronicled and thankfully need little recapitulation here, it is worth saying something about the circumstances that underpinned their rise. While Mao's and Stalin's ascent to power may not have had strictly populist origins, their consolidation of power, their utilization of the familiar tropes of national reconstruction, their lack of accountability and the absence of effective political opposition are all reminders of why some form of democracy remains desirable—despite all of the shortcomings noted in earlier chapters. More pointedly for the purposes of this discussion, Russian and Chinese history are painful reminders of the violent, unpredictable changes that the traumas of internal conflict and crisis can unleash. Likewise, it is important to remember that Hitler's rise and popularity owed much to the economic catastrophe that unfolded in Germany as a direct consequence of the misguided policies of the victorious allies in the aftermath of World War I.[11]

The potentially disastrous consequences of the allies' demand for war reparations were foreseen by the most brilliant economist of his time: John Maynard Keynes.[12] The fact that Keynes' analysis was initially ignored is another illustration of the possibility that the ideas of experts may not find a receptive audience simply because they are accurate, and are often neglected until it is too late.[13] The claim that economic circumstances can have a profound impact on political and social outcomes looks relatively uncontroversial these days though, even if policymakers are prone to forgetting quite how crises can develop and what their political impact can be.[14] This is especially the case in the US, where successive economic catastrophes have sparked populist responses, some of which have been surprisingly progressive.

Although the Trump administration and its supporters may not be an encouraging advertisement for the possibility of a progressive form of populism, this does not mean that that American populism is inevitably reactionary or right-wing. Not only did what we now think of as populism actually originate in the US, but its first incarnation was also striking for its radical, anti-establishment, egalitarian foundations. Michael Kazin describes American populism as a 'persistent yet mutable style of political rhetoric with roots deep in the nineteenth century', and while it may have gone through a number of changes that reflect shifting, contingent realities, its origins can be found in the collective actions of farmers and workers as they revolted against the 'money power' of the plutocracy.[15] The People's Party that emerged from this febrile atmosphere won seats in

House of Representatives and the Senate in the elections of 1894, and seemed set to challenge the Democrats as the second party.[16]

This early populist efflorescence was cut short, however, not least by its inability to come to terms with America's domestic racial politics and the 'negro question'.[17] The point to emphasize, however, is that America's working class, and even its usually highly conservative rural sector, have not only been united at times, but they have also supported surprisingly radical economic agendas and a more powerful role for the state.[18] Much has changed in the US since this period, of course, but many observers have pointed to the fact that economic crises of a sort that gripped late nineteenth-century America—and which have reappeared with all-too-predictable regularity ever since—have the potential to act as critical junctures, potentially transforming the conventional wisdom about policy and much else.[19] It is worth remembering that populist sentiment and political pressure was instrumental in bringing about the original New Deal that emerged from the Great Depression, as well as the subsequent civil rights movement of the 1950s and 1960s.[20] Progressive politics and populism are not necessarily or inevitably antithetical.

Nor, of course, are they synonymous or necessarily synergistic. Hitler's rise was facilitated by the destruction of 'the faith of the people in the economic structure of German society',[21] as the life savings of the middle and the working classes were wiped out by catastrophic hyperinflation. As Keynes also pointed out, 'there is no subtler, no surer means of overturning the existing basis of society than to debauch the currency.'[22] Economic crises still have the capacity to galvanize populist reaction, especially when they highlight the differential impact of crises on different strata of society. As we saw in Chap. 3, growing inequality is becoming endemic across much of the world. Significantly, it is increasing more quickly *within* rather than between nations, another trend in which the US is ahead of the economic curve. Equally importantly, perhaps, as Thomas Piketty pointed out in his best-selling analysis of the crisis of contemporary capitalism, 'there is absolutely no doubt that the increase in inequality in the US contributed to the nation's financial instability'.[23]

The Political Economy of Populism

The intention here is not to offer yet another analysis of the global financial crisis that trashed the reputation and standing of American capitalism in 2008,[24] but to emphasize the importance of underlying structural change, the dangers of economic mismanagement and its potential to spill over into potentially 'bad' populism.[25] This is especially true when it is widely perceived that the international regulatory environment had been recklessly reconfigured to reflect the preferences of vested interests to benefit some more than others.[26] What is worth emphasizing in the context of a discussion of the possible driving forces of populism is that economic crises can shine a light on the cronyistic, self-serving relationships that have come to characterize contemporary capitalism everywhere it seems.[27] Even the People's Republic of China is not immune,[28] and this has potentially even greater consequence for a regime that is supposed to represent the will and the interests of the proletariat, not the plutocracy.

In China, the CCP has largely managed to control popular unrest over economic and environmental issues,[29] even though protests about both are on the rise.[30] Once again, however, it is the US that has been most affected by a shift in collective consciousness as more people recognize that their interests may not be synonymous with the '1 percent' that have disproportionately benefited from the new economic order and destabilized its foundations in the process. John Judis usefully highlights the fact that the excesses and failures of neoliberalism have been attacked from the Left *and* the Right, in the US: while the 'Occupy Wall Street' movement may have garnered the most attention in the aftermath of the global financial crisis (GFC), conservative forces have long railed against various aspects of the neoliberal project, especially the role globally oriented elites have supposedly played in undermining the national interest. Populists such as Ross Perot, Pat Buchan and more recently the Tea Party articulated a liberal, anti-establishment agenda that revealed deep divisions among America's conservative political forces.[31]

The point of this digression on America's domestic political economy is to highlight, once again, the importance of structural change and public policy in determining political alignments and perceptions about what government priorities ought to be. Many conservatives are not hostile to climate change mitigation simply because they are skeptical about the need to do something about it—although many of them are that, too, of course—but because they don't regard it as a first order issue. In this they have some significant if controversial support from the likes of the Danish political activist Bjørn Lomborg, who argues that scarce resources and money should be devoted to immediate problems rather than uncertain future outcomes.[32] Whatever one thinks of Lomborg's work—and it has been subjected to some withering critiques[33]—he does demonstrate how a well-focused and well-funded 'ideas broker' and/or 'policy entrepreneur' can make an impact, especially when they potentially give credibility to powerful vested interests that might otherwise be badly affected by new forms of regulation.[34] It is striking that the ideas of prominent Australian policy entrepreneur, Ross Garnaut, were much less enthusiastically received when he talked about climate change mitigation than they were when he articulated a broadly pro-business reform agenda in an earlier policy incarnation.[35]

For those attempting to articulate an alternative policy agenda, even in the somewhat less daunting arena of economic reform, these are important considerations. The good news is that there are historical examples of progressive public policy reform as a consequence of profound economic dislocation: the New Deal and the postwar reconstruction of large parts of the international economy are important examples of this possibility.[36] It is partly as a consequence of the systematic dismantling of the regulatory framework that was developed in this period, and which kept the speculative excess of the private sector in check, that new crises developed with all too predictable inevitability.[37] The challenge for those trying to highlight the need for climate change mitigation is to make the case for the need and potential benefits of similarly systemic forms of regulation.

Of the many obstacles facing such an objective, not the least is the fact regulation is something of an anathema to many, and not just in the sectors of the business community that are likely to be affected by any shift in the policy regime. While it has been frequently pointed out that there is actually no such thing as a deregulated economy[38]—any economic system needs the state to provide the rules that allow commerce to occur and private property to be protected—this has not stopped powerful vested interests from lobbying against regulations that may affect their profitability. Unfortunately, this is part of the reason that effective policy to combat climate change has proved so difficult to achieve, especially in the US.[39] The potential incompatibility of private and collective interests is well summarized by David Rothkopf, who suggests that if we

> [a]sk why the world can't or won't address concerns from global warming, to embracing new forms of energy, to containing global diseases, to regulating global derivatives markets, and you will see the not-so-visible hand of the megaplayers. These enduring private actors, which were built on an idea once created to serve national interests—the idea of the corporation—have morphed into a group that plays a leading role shaping national and international priorities often without regard for the interests of any one society

anywhere—a group with the money and power to institutionalize their ideologies and serve their interests by successfully supporting efforts to translate their ideas into laws or, alternatively, carefully carve out legal and regulatory voids.[40]

Yet many in the private sector, especially in the Anglo-American economies, remain wedded to idea that 'the state' is part of the problem when it comes to economic development and decisions. However, as we saw in Chap. 3, there are signs that at least some business leaders recognize the need for reform, and that at least some of them are championing the need to put a price on carbon, for example. Importantly, so do organizations such as the World Bank, which have the capacity to influence the adoption of climate friendly policies in the developing economies.[41]

THE CONTRADICTIONS OF POPULISM

Encouraging as some of these developments may be, there is a noteworthy disjuncture between the sort of populist upsurge that occurred in the aftermath of the GFC in particular, and the demand for change some of these movements tried to encourage. While the Occupy Wall Street (OWS) movement, and even the recent *gilets jaunes* protests in France, may be motivated by unhappiness with the economic system, they have been less good at building on that initial momentum to develop alternative strategies and ideas about what might be done. One important exception to this trend are the so-called Indignados and the associated 15-M Movement in Spain, which began as a form of 'leaderless revolution',[42] in response to Spain's very high levels of unemployment, but which rapidly evolved into a very effective political party: Podemos has already become the third largest political party in the Spanish parliament and has nearly half a million members. Interestingly, Podemos explicitly draws on the writing of Laclau and Mouffe for their inspiration.[43]

Yet there is generally still a fundamental and potentially fatal disjuncture between the sorts of time frames in which protestors want to see action of some sort being taken, and the capacity of even well-intentioned states to react, or to overcome powerful sources of opposition to reform. For all the attention that has been paid to the Occupy Wall Street (OWS) movement's creative use of new forms of social media and networking,[44] for example, any movement of this sort faces a more formidable set of challenges that are summarized by Roberts:

opponents of neoliberalism must acknowledge three points. The first is that an opposition movement must have the capacity to coordinate and control action, which OWS clearly lacked. Second, it must have a philosophy of action that concedes the possibility of tactical alliances with other social actors. And third, it must have an overarching view about the role of state and economy—a new paradigm—that explains in a concise and appealing way its alternative to the status quo and forestalls unending debate about what the demands of the movement should be.[45]

This is a formidable challenge within the national framework of traditional Westphalian polities. It is even more complex when trying to generate new ways of thinking about and responding to problems that transcend national borders. Indeed, as Nancy Fraser points out, 'far from institutionalizing debate among citizens who share a common status as political equals, post-Westphalian publicity (sic) appears in the eyes of many observers to empower transnational elites, who possess the material and symbolic prerequisites for global networking.'[46] In other words, even if a transnational public sphere is emerging, which seems uncontroversial at this point, it is still dominated by the sorts of forces that have tended to shape policy debates and implementation at the national level. This is another reason for thinking that transforming the *domestic* debate and context, especially in those countries where this is at least feasible in practice, is a necessary part of meaningful change.

In this regard, at least, the evidence is somewhat encouraging, not least for those who think that some form of democratic political process is more likely to produce sustainable results and is, in any case, normatively preferable. One expression of this potential, and the possibility that ordinary people can both reclaim agency and become politically engaged is the so-called movement of the squares, or the sudden upsurge in protest and political activism that was epitomized by the Arab Spring and which led to similar protest movements in the US, Spain, Greece and elsewhere. Although there are many reasons for feeling deflated about some of these movements in retrospect, there is one important and encouraging legacy of this period that is important to keep in mind for anyone hoping for large-scale transformative social action intended to 'save the planet': activism is possible and it can have a major impact, not least on the activists themselves.

As Paolo Gerbaudo argues in his important (and rather inspiring) book, *The Mask and the Flag*, one of the longer-term consequences of these populist upsurges has been to establish—in some minds, at least—the idea of what he calls 'citizenism', or 'the ideology of the "indignant citizen"'. Within this emerging ideology, he argues, '"the citizen", or "the citizenry", as a sort of libertarian and individualist variation on the subject of populism's "the People", has be a more participatory and bottom-up approach to the populist demand for popular sovereignty, the new revolutionary object'.[47] If we accept that political participation is a key part of achieving effective environmental policies,[48] and that democracies are more likely to generate such policies in the first place,[49] the existence of an active, engaged, informed citizen is potentially important. It is at this level of 'everyday politics', Boyte argues, that some forms of populism can actually be productive:

> politics needs to be owned by the people, not professionals. It is rooted in everyday settings, not in government or elections or mobilization campaigns. It is thus populist, in the sense that Thomas Jefferson once argued, resting on the premise that the only safe repository of the powers of the society are the people themselves.[50]

Given that the Arab Spring has fizzled out and that political repression and/or dysfunction seems as endemic as ever in much of the Middle East,[51] the prospects for everyday politics do not look bright, perhaps. Likewise, Greece's populist government has had to radically scale back its ambitions in the face of an EU-imposed austerity program its government appeared to have little option other than to accept.[52] Greece is also a reminder of one other climate-related possible obstacle to transnational action and national political consensus: migration.

THE MIGRATION CRISIS

There is little doubt that the development of 'bad' populism and the re-emergence of far-right politics in parts of Europe in particular is a consequence—in part, at least—of anxiety about migration, illegal and otherwise.[53] Greece has had its own version of this phenomenon in the ultranationalist 'Golden Dawn' party, but so has Germany, with the emergence of the 'Alternative for Germany' (AFD). As we saw in Chap. 4, even the progressive Scandinavians have not been immune from a backlash against large-scale migration and the possible problems this may bring in its wake.[54] The first point to make is that even if such concerns are overblown, the product of ignorance and/or racism, or even normatively indefensible, that does not make them any less real as far as large numbers of people are concerned. It also makes them a potential obstacle to the creation of a broadly based progressive political agenda of a sort that would seem necessary to address climate change and much else.

The second point to make is that the 'immigration crisis', if that is the right way to describe it, is already enormous and likely to get much worse as a direct result of climate change. There are already an estimated 68.5 million 'forcibly displaced' people in the world, of whom more than 85 percent live in developing countries, which are often not well placed to provide refuge.[55] There is a plausible argument that many of these people are directly or indirectly fleeing the consequences of climate-induced crises that have evolved into military conflicts; Syria is perhaps the most consequential case in point.[56] But even if people are only escaping 'normal' environmental problems of flood or famine, their numbers are likely to grow dramatically as a deteriorating natural environment forces people from their homes. No one knows exactly how many, of course, but by some estimates there may already be 25 million climate refugees, and the total number could reach something like 250 million by 2050.[57] Immigration is perhaps the most intractable, politically and ethically challenging aspect of responses to climate change, populist or otherwise. As such, it merits serious discussion.

Even if we put aside the issue of whether all of the people currently trying to gain access to privileged parts of the planet such as the US, Australia or the EU are 'genuinely' fleeing persecution or environmental degradation, the sheer number of migrants raises potentially enormous practical problems and ethical issues, not the least of which is its potential impact on domestic and international politics. As Paul Collier notes, one of the most challenging, confronting and consequential aspects of the current crisis is the growing realization and/or determination that 'migration cannot be left to the decisions of individual migrants; it must be managed by governments'.[58]

While migration may have been a feature of the human experience from the time we collectively learned to walk upright, this is unlikely to prove a compelling argument for those who feel displaced, threatened, and disempowered by what they—rightly or wrongly—perceive to be a large influx of people with different values, religions and customs. The fact that they often look different is plainly an issue for some, too. It has also been frequently pointed out that some of the most fervent right-wing populist supporters are often from working class, deindustrializing communities who feel threatened by the sort of social change migrants seeking work and affordable accommodation often bring. The main interaction of new migrants with cosmopolitan elites, by contrast, may be in their role as members of the rich world's rapidly expanding servant class.[59] As Eric Kaufmann points out in his important and discomfiting book, 'the disjuncture between official liberal cosmopolitanism and local-level tribalism is glaring'.[60]

No doubt this will be regarded as something of a cheap shot by some readers, but there is an even more important and practical reason for giving serious consideration to some of the—profoundly unsettling and not easily resolved—problems large-scale immigration can raise. If the potential host country has the capacity to take its responsibilities to the new arrivals seriously—which in the best of all possible worlds it should, of course—then it will need to spend a lot of time and money turning them into productive members of society.[61] This is likely to be an expensive, labor-intensive, long-term commitment, and a huge challenge with large numbers of people. This is not to say that such an effort should not be made. It does mean that there may be real practical limits to the numbers of new arrivals that any country can accommodate if social harmony is to be maintained and the arguments of the racists and the marginalized are to be countered.

This is why the numbers are so important. Unless you subscribe to the belief that anyone should be able to go anywhere they wish, simply because they are human, and that national borders are an artificial, arbitrary imposition on our universal human rights, then someone has to decide who gets to move and under what circumstances. Universal cosmopolitanism is a perfectly legitimate position to take, especially if you are an academic with no responsibility for enacting policy,[62] but it's not one that is likely to get you elected as most politicians recognize.[63] Although intergovernmental agencies such as the UN may play a role in deciding the fate of refugees at the margins, ultimately it is still states that determine—or try to determine—who is allowed into a particular part of the planet, no matter how arbitrary and contingent that political space may be. In this context, it is significant that some would-be refugees and migrants are trying to escape from state failure or from the competition for the control of the state by rival elites.[64] And who can blame them? I know I would.

Plainly, a rich country with a record of successful immigration, such as Australia—or the US for that matter—could accommodate more refugees than it currently does. But how many should that be? Australia is a country of 25 million people, so would 10 percent be fair? Should it be more, and solely dependent on need? Over what time period should this happen? What impact would such a very large, possibly rapid inflow of new arrivals have on the existing population, especially if they came from very different social backgrounds? In this context, should Australia's much criticized approach to illegal migration and its 'stop the boats' mantra be praised or vilified? There is little doubt that successive conservative governments in Australia have successfully accomplished what they set out to do, albeit at the cost of Australia's international reputation, to say nothing of the mental health and human rights of the people rotting away in 'off-shore processing centres'.[65]

There is no unambiguously right answer to these sorts of questions, either ethically or pragmatically,[66] but one thing that needs to be considered is the impact that large, uncontrolled, unsolicited flows of migrants are likely to have on the existing population and national politics. At the very least, it would not be unreasonable to assume that not everyone would welcome such a dramatic change in the makeup of the population and the additional burden that would place on social infrastructure. Donald Trump rose to power partly on the idea that uncontrolled, illegal migration via Mexico was a threat to the US. Even if some of his claims are inaccurate, false or quite simply preposterous, the fact that they resonate with enough people to get him elected reminds us of what a powerful and divisive influence the politics of immigration can be.[67] Even Australia has its own, low-wattage version of racist populism in the form of Pauline Hanson.[68]

The significance of the immigration debate in the context of efforts to create a broadly based consensus on the need for urgent action on climate change is that it threatens to undermine the possibility of any national agreement on policy priorities. As we have seen, large-scale flows of illegal, undocumented and—to speak plainly—often unwanted migrants have transformed the political landscape in Europe.[69] In such a context, it would be unsurprising if environmental issues, which unfold in a comparatively longer time-frame, were pushed down the policy agenda. That is, of course, precisely what is happening as policymakers in even long-standing democracies grapple with even more pressing problems of political survival and lurch from one crisis to another—pursued by bad populists who criticize their every move.

The implications of this discussion are sobering to say the least, but frequently ignored, underestimated or dismissed with Panglossian assertions such as 'we can do it', to quote the well-intentioned Angela Merkel. Plainly not even Germany, with all its wealth, capacity and historical need to demonstrate international good citizenship, could rapidly assimilate a million migrants.[70] If good intentions had continued to triumph over a clear-eyed calculation of national capabilities and vulnerabilities—including the sentiments of the established citizenry—the stability of the political system in Germany and many other places might have been threatened. Even more importantly from the perspective of long-term transnational cooperation, the migration crisis may ultimately destroy the EU[71]—the only real and enduring example we have of highly institutionalized and effective cross border cooperation. The practical and symbolic impact of the EU collapsing under the weight of some of its own internal contradictions would be profound, not least for the very idea of effective international collective action.[72]

The federal system of the US and its famous system of checks and balances is also currently being put to its most searching test in modern history; not everyone is convinced that it will survive.[73] The bitter feud over Trump's wall and its toxic, polarizing impact on American politics reminds us that no country is immune to the stresses that migration and threats to traditional identities seem to pose. Likewise, the country of my birth has embarked on one of the most traumatic, destructive periods and processes in its entire history because many people were unhappy not only about Britain's relationship with the EU, but with a perceived loss of sovereignty and control of national borders.[74] The people who hold such views don't have to be correct, nor do the policies they espouse have to make sense, for them to pose a profound challenge to the prevailing political order. The chances of anyone paying much attention to the *real* challenge we collectively face and cannot escape are not good in such circumstances.

The principal conclusion that flows from all of the above is not a happy one, but one we cannot avoid: unless the national political context is relatively stable and has the luxury of focusing on long-term problems as well as the more immediately visible variety, there is little hope of effective collective action at the international level. To be sure, it's possible to address more than one problem at a time, but it is much easier to kick problematic polices into the long grass—or the hands of future generations who will have little option about whether to act or not. The only question will be how and under what circumstances they will attempt to manage the abdication of responsibility on the part of the current generation of policymakers. This is why, I think, those countries that still have the capacity, the opportunity and even the time to act, must do so and hope that their efforts will inspire other people, especially outside the traditional halls of power, to demand similar actions from their own policymakers and leaders. If we hope to effect change, we must start where we live, one government at a time.

Lamentations of the Lucky Country

There are few countries better placed—or with more compelling reasons—to take climate change seriously than Australia. Australia is, after all, the driest continent on earth, still a major agricultural producer, and remarkably free of many of the debilitating questions of fundamental insecurity that afflict so many other parts of the world. If a country that has just experienced one of its hottest years on record can't summon up the collective will to do something about global warming,[75] who can? Self-interest doesn't get much more compelling, one might think, but this has not stopped political leaders and powerful vested interests form doing everything they can from discouraging effective action being taken.

No doubt we all think the countries we happen to live in are special is some way. Thankfully, Australians are generally not prone to delusions of grandeur, laboring under the illusion that they are on a mission from God, or convinced that they are the embodiment of a uniquely important civilization. But there are some features of the country that are at least noteworthy. No other population enjoys an entire continent to itself, for example. True, most of it is judged uninhabitable by the vast majority of the population who cram into a handful of big cities on the coast, but it ought to engender a sense of security and good fortune.[76] But when the prominent Australian intellectual, Donald Horne, described Australia as the 'lucky country' in 1964 he was, in fact, being ironic: Australians had the great misfortune of being governed by 'second rate' mediocrities who couldn't help but do moderately well given the country's natural advantages and resource base.

Much has changed in Australia in the half century or so since Horne wrote so witheringly about Australia's stunted cultural life and national underachievement. The country and its political elites now receive surprisingly gushing endorsements from the likes of *The Economist*.[77] There is still a lively debate within Australia about whether such accolades are actually justified,[78] but whatever one thinks about Australia's economic management, effective environmental policy has proved to be beyond capabilities of the two major political parties. Indeed, despite setting a world record for uninterrupted economic growth, Australia has had six prime ministers in 11 years, which may also be some sort of record, and which has led to unflattering comparisons with banana republics.[79] Although there are contingent problems with the design and operation of political institutions in Australia, it is noteworthy that climate change policy has played a noteworthy role in the downfall of a number of leaders.

I shall not test the patience and attention spans of non-Australian readers by providing an exhaustive description of climate politics in my adopted home,[80] but this country does highlight a few issues of wider significance. Former Australian Labor Party (ALP) leader Kevin Rudd was possibly the most ambitious prime minster Australia has ever had, but his downfall highlights the perils politicians face in addressing what Rudd famously called the 'greatest moral challenge of our time'.[81] His subsequent retreat from his lofty rhetoric in the face of what prominent Australian journalist Paul Kelly called the 'best special-interest campaign in recent decades',[82] contributed to his defeat at the polls and demonstrated the power of the mining sector in Australian politics.[83] In some ways, this outcome and the well organized opposition of the coal lobby in particular should have come as no surprise: such forces have exercised a major influence on 'Australia's environmental policies since they became entrenched under the Liberal leader John Howard during the 1990s.[84] Indeed, in many ways, climate change policy has been even more problematic for the conservative, pro-business Liberal Party than it has for the ALP. The Liberal's most progressive and well-informed leader ever, Malcolm Turnbull, lost the leadership of his party on two separate occasions, in part because he attempted to develop a more sustainable approach to energy and climate policies.[85]

Part of Australia's—and America's, of course—difficulty in developing policies that could prove effective in countering global warming stem from the idiosyncrasies of its historical political development. On the plus side, Australia has compulsory voting and so there is a potentially engaged and even informed electorate. On the negative side of the ledger, however, Australia suffers from a hopelessly truncated electoral cycle and an anachronistic federal system, replete with a highly unrepresentative Senate.[86] This has allowed people to game the system and ride on the coattails of people like Pauline Hanson to get elected, often with a ludicrously small share of the vote.[87] Problematically enough for mainstream politicians, 'the minor party vote increased most strongly in the period in which wages and income growth was most healthy'.[88] Consequently, Australia's political class is locked in a relentless cycle of heightened electoral sensitivity and combativeness, exacerbated by a 24-hour news cycle. The limited prospects for effective leadership and the disillusionment of Australian voters as a consequence are sadly all too predictable.[89]

Australia has a structurally powerful business sector that, until relatively recently, has campaigned relentlessly against government 'interference' in the market place and associated investment decisions.[90] In addition, Australia's only national 'quality' broadsheet, *The Australian*, has, according to Robert Manne, conducted a prolonged and intellectually incoherent campaign against action on climate change, which has undermined the hold in public life of the central values of the Enlightenment, Science and Reason.'[91] In the post-Rudd and Turnbull era, skepticism within the Liberal-National Party coalition increased regarding the value of multilateral action, and the current government of Scott Morrison remains firmly committed to assisting the coal industry, despite the dire and highly plausible warnings of the IPCC.[92] State governments are equally hostage to parochial political concerns, as the Queensland government's support for the notorious Adani group's mega-coal project demonstrates.[93]

And yet there are some encouraging signs. Australia has a dedicated Green party, a number of active civil society organizations that take climate change seriously, and there are high levels of public support for effective action on climate. According to a recent survey '73% (up from 66% in 2017) of respondents are concerned about climate change, and a clear majority, 68%, believes the government should set domestic targets to comply with our Paris commitments'. Equally strikingly, 67 percent want coal-fired power to be phased out within 20 years.[94] More generally, there is evidence that, despite a degree of cynicism and apathy on the part of voters, given issues and ideas that they consider to be potentially important and relevant, they are prepared to engage with the political system.[95] There are also meaningful empirical examples of voter (re)engagement in the political process actually having a transformative impact, albeit on a small scale. Hendricks suggests in this context that 'the path to reconnecting citizens to their democratic systems may not lie in specific institutional designs but in citizens themselves'.[96] Simply taking part in the political process in the way many advocates of deliberative democracy emphasize is in some ways just as important as the eventual outcome, it would seem.

While this local-level activism looks like the antithesis of populism—especially the thin variety—in some respects it is a reminder of the potential importance of agency and the challenge of scaling up political responses from the electorate to the national polity, and ultimately to each country's external relations. Yes, it is a dreadfully labor intensive process and possibly not applicable in every context, but as the old Chinese proverb has it, a journey of a thousand miles begins with a single step.[97]

Conclusion

This chapter has covered a lot of ground. Once again, it is easier to describe the problems than the solutions. There is little value in ignoring the obstacles that confront would-be reformers, however. On the contrary, if anything is to be done to confront the 'greatest moral challenge of our time', then it is highly likely that it will require new ways of thinking about and responding to what is an unprecedented, interconnected series of challenges and problems. The implications of this possibility are taken up in the final chapter. What this chapter has suggested is that there are, indeed, some grounds for cautious optimism and some modestly encouraging signs. The 'strike' by schoolchildren over government inaction in the face of climate change in Australia was one such, even though it was decried by all the usual suspects.[98] Of the many hilarious and/or pointed signs carried by the young protestors, one in particular captured the spirit of the day and the yawning intergenerational gulf: 'we'll be less activist if you'll be less shit.' Quite so.

As we have seen, the capacity to mobilize around progressive ideas and policies already exists, even if there are still formidable obstacles to scaling them up to the necessary transnational levels. Significantly, it is not only the young and disaffected who understand the possible implications that uncontrolled climate change is likely to have on their lives. There is a growing recognition by at least some members of the 'business community', and *some* members of the political elite, that determines the political responses of individual nations that urgent action needs to be taken. The fact that such responses are not only relatively spontaneous protests with short shelf lives, and that some of the basic concerns are gradually permeating the consciousness of a wider cross section of society, is a hopeful sign. The question, as ever, is whether what Boyte calls the 'stirrings of a new politics' can have a significant impact in the time available for effective action.[99] This is where populist pressure and participation can potentially make a difference.

Notes

1. Thaxton, R. (2008) *Catastrophe and Contention in Rural China: Mao's Great Leap Forward Famine and the Origins of Righteous Resistance in Da Fo Village* (Cambridge: Cambridge University Press); Conquest, R. (2018) *The Great Terror: Stalin's Purge of the Thirties* (New York: Random House).
2. Chan, M. (2017) 'Disruptive, intolerant and populist icon: why Trump is seen as an American Mao', *South China Morning Post*, January 20; Brandenberger, D. (2010) 'Stalin's populism and the accidental creation of Russian national identity', *Nationalities Papers* 38(5): 723–739.
3. Filipov, D. (2017) 'For Russians, Stalin is the "most outstanding" figure in world history, followed by Putin', *Washington Post*, June 26; Anderlini, J. (2016) 'The return of Mao: a new threat to China's politics'. *Financial Times*, September 29.
4. Babones, S. (2017) 'Xi Jinping: Communist China's first populist president', *Forbes*, October 20.
5. Nathan, A.J. (2018) 'China: Back to the future'. *New York Review of Books* May 10; Weiss, J.C. (2014) *Powerful patriots: nationalist protest in China's foreign relations* (Oxford: Oxford University Press). But also see, Johnston, A.I. (2017) 'Is Chinese nationalism rising? Evidence from Beijing', *International Security* 41(3): 7–43.
6. Krugman, P. (2018) 'Stop calling Trump a populist,' *New York Times*, August 2.
7. Mounk, Y. (2018) *The People Vs. Democracy: Why Our Freedom is in Danger and how to Save it* (Cambridge, Mass.: Harvard University Press); Rehman, I. (2017) 'Rise of the reactionaries: The American far right and U.S. foreign policy', *The Washington Quarterly* 40(4): 29–48.
8. Mudde, C. (2004) 'The populist zeitgeist', *Government and Opposition*, 39(4), p. 544.
9. Gellately, R. (2002) *Backing Hitler: Consent and Coercion in Nazi Germany* (Oxford: Oxford University Press).
10. Su, Y. (2011) *Collective Killings in Rural China During the Cultural Revolution* (Cambridge: Cambridge University Press).
11. Kershaw, I. (2014) *Hitler: A Biography* (London: Routledge).
12. Keynes, J.M. ([1920] 2004) *The Economic Consequences of the Peace* (New York: Dover).
13. Blyth, M. (2002) *Great Transformations: Economic Ideas and Institutional Change in the Twentieth Century* (Cambridge: Cambridge University Press); Legro, J.W. (2005) *Rethinking the World: Great Power Strategies and International Order* (Ithaca: Cornell University Press).
14. Reinhart, C.M. and Rogoff, K., S. (2009) *This Time Is Different: Eight Centuries of Financial Folly* (Princeton: Princeton University Press).

15. Kazin, M. (1998) *The Populist Persuasion: An American History* (Ithaca: Cornell University Press), p. 5.
16. Judis, J.B. (2016) *The Populist Explosion: How the Great Recession Transformed American and European Politics* (New York: Columbia Global Reports).
17. Jordan, W.D. (2013) *White Over Black: American Attitudes Toward the Negro, 1550–1812* (Chapel Hill: UNC Press).
18. Some argue that this is precisely what needs to happen now as an antidote to the disintegrative impact of identity politics. See, Ganesh, J. (2019) 'Time for America to embrace the class struggle'. *Financial Times*, January 16.
19. Funke, M., Schularick, M. and Trebesch, C. (2016) 'Going to extremes: Politics after financial crises, 1870–2014', *European Economic Review* 88, 227–260.
20. Boyte, H.C. (2010) *Everyday Politics: Reconnecting Citizens and Public Life* (Philadelphia: University of Pennsylvania Press), p. 21.
21. Shirer, W.L. (1991) *The Rise and Fall of the Third Reich: A History of Nazi Germany* (London: Pan Books), p. 86.
22. Keynes, *op cit*, p. 235.
23. Piketty, T. (2014) *Capital in the Twenty-first Century* (Cambridge, Mass.: Belknap Press), p. 297.
24. See, Bell, S. and Hindmoor, A. (2015) *Masters of the Universe, Slaves of the Market* (Harvard: Harvard University Press); Morris, C.R. (2008) *The Trillion Dollar Meltdown: Easy Money, High Rollers, and the Great Credit Crash* (London: Public Affairs).
25. Eichengreen, B. (2018) *The Populist Temptation: Economic Grievance and Political Reaction in the Modern Era* (Oxford: Oxford University Press).
26. Tett, G. (2009) *Fool's Gold: How Unrestrained Greed Corrupted a Dream, Shattered Global Markets and Unleashed a Catastrophe* (London: Little, Brown).
27. Culpepper, P.D. and Reinke, R. (2014) 'Structural power and bank bailouts in the United Kingdom and the United States', *Politics & Society* 42(4): 427–454; Hacker, J.S. and Pierson, P. (2017) *American Amnesia: How the War on Government Led Us to Forget What Made America Prosper* (New York: Simon and Schuster).
28. Pei, M. (2016) *China's Crony Capitalism* (Harvard University Press).
29. Stern, R.E. and Hassid, J. (2012) 'Amplifying silence: Uncertainty and control parables in contemporary China', *Comparative Political Studies* 45(10): 1230–1254.
30. Browne, A. (2017) 'The underclass that threatens Xi's "China Dream"'. *Wall Street Journal* December 5; Babones, S. (2017) 'Red Alert for China's pollution protesters', *Aljazeera*, February 20.

31. Judis, J.B. (2016) *The Populist Explosion: How the Great Recession Transformed American and European Politics* (New York: Columbia Global Reports).
32. Lomborg, B. (2003) *The Skeptical Environmentalist: Measuring the Real State of the World*, vol. 1 (Cambridge: Cambridge University Press).
33. Karp, P. (2017) 'Experts reject Bjørn Lomborg's view on 2C warming target', *The Guardian*, May 21.
34. Mintrom, M. and Norman, P. (2009) 'Policy entrepreneurship and policy change', *The Policy Studies Journal* 37(4): 649–667.
35. Beeson, M. and Stone, D. (2013) 'The changing fortunes of a policy entrepreneur: The case of Ross Garnaut', *Australian Journal of Political Science* 48(1): 1–14.
36. Frieden, J.A. (2006) *Global capitalism: Its Fall and Rise in the Twentieth Century* (New York: W.W. Norton).
37. Crotty, J. (2009) 'Structural causes of the global financial crisis: a critical assessment of the 'new financial architecture'', *Cambridge Journal of Economics* 33(4): 563–580.
38. Cerny, P.G. (1993) 'The deregulation and re-regulation of financial markets in a more open world', in, P.G. Cerny *Finance and World Politics: Markets, Regimes and States in the Post-hegemonic Era* (Aldershot: Edward Elgar): 51–85; King, R. (2007) *The Regulatory State in an Age of Governance: Soft Words and Big Sticks* (Basingstoke: Palgrave).
39. Hein, J.E. and Jenkins, J.C. (2017) 'Why does the United States lack a global warming policy? The corporate inner circle versus public interest sector elites', *Environmental Politics* 26(1): 97–117.
40. Rothkopf, D. (2012) *Power, Inc: The Epic Rivalry Between Big Business and Government – and the Reckoning That Lies Ahead* (New York: Farrar, Straus and Giroux), p. 18.
41. World Bank (2016) *Climate Change Action Plan, 2016–2020* (Washington: World Bank Group).
42. Ross, C. (2011) *The Leaderless Revolution: How Ordinary People Will Take Power and Change Politics in the 21st century* (London: Simon and Schuster).
43. Kioupkiolis, A. (2016) 'Podemos: the ambiguous promises of left-wing populism in contemporary Spain', *Journal of Political Ideologies* 21(2): 99–120.
44. Costanza-Chock, S. (2012) 'Mic check! Media cultures and the Occupy movement', *Social Movement Studies*, *11*(3–4), 375–385.
45. Roberts, A. S. (2012). 'Why the Occupy movement failed', *Public Administration Review* (September/October), p. 761.
46. Fraser, N. (2007) 'Transnationalizing the public sphere', *Theory, Culture & Society*, *24*(4), p. 16.

47. Gerbaudo, P. (2017) *The Mask and the Flag: Populism, Citizenism, and Global Protest* (London: Hurst), p. 8.
48. Stevenson, H. and Dryzek, J.S. (2014) *Democratizing Global Climate Governance* (Cambridge: Cambridge University Press).
49. Neumayer, E. (2002) 'Do democracies exhibit stronger international environmental commitment? A cross-country analysis', *Journal of Peace Research* 39(2): 139–164.
50. Boyte, *op cit*, p. 4.
51. Kadri, A. (ed. 2016) *Development Challenges and Solutions after the Arab Spring*. (Basingstoke: Palgrave).
52. Rogers, C. and Vasilopoulou, S. (2012) 'Making sense of Greek austerity', *The Political Quarterly* 83(4): 777–785.
53. Galston, W.A. (2018) *Anti-Pluralism: The Real Populist Threat to Liberal Democracy* (Yale: Yale University Press).
54. Ekman, I. 2019. 'Swedish unexceptionalism: Sweden's elections show that a strong far-right is Europe's new normal.' *Foreign Affairs*, September 10.
55. UNHCR (2018) *Statistical Yearbook 2016*, 16th Edition (UN Refugee Agency), https://www.unhcr.org/figures-at-a-glance.html
56. Gleick, P.H. (2014) 'Water, drought, climate change, and conflict in Syria', *Weather, Climate, and Society* 6(3): 331–340.
57. Myers, N. (2002) 'Environmental refugees: a growing phenomenon of the twenty-first century', *Philosophical Transactions: Biological Sciences* 357(1420): 609–613.
58. Collier, P. (2013) *Exodus: How Migration is Changing Our World* (Oxford: Oxford University Press), p. 255.
59. *The Economist* (2011) 'The servant problem', December 17; Doherty, B. (2016) 'Revealed: the systematic exploitation of migrant workers in Australia', *The Guardian*, October 29.
60. Kaufmann, E. (2018) *Whiteshift: Populism, Immigration and the Future of White Majorities* (London: Allen Lane), p. 51.
61. Trying to work out how much it costs to resettle refugees and/or asylum seekers is far from straight forward. Estimates in the US are in the vicinity of US$65,000 per year. If they are to become productive citizens, of course, they may need English tuition and training on top of this. Valverde, M. (2017) 'Estimating the costs of refugee resettlement in the United States, assistance in home region', *Politifact*, September 21. https://www.politifact.com/truth-o-meter/article/2017/sep/21/refugee-resettlement-united-states-versus-assistan/. In Germany, the government was estimated to have spent 5.3 billion euros on social welfare, and another 21 billion on 'refugee related expenditure'. The intention here is not to criticize or denigrate, simply to point out that good intentions can be expensive and resented by those who feel it is occurring at their expense. See,

Trines, S. (2017) 'Lessons from Germany's refugee crisis: Integration, costs, and benefits', *World Education News & Reviews*, May 2. https://wenr.wes.org/2017/05/lessons-germanys-refugee-crisis-integration-costs-benefits

62. Archibugi, D. (2000) 'Cosmopolitical democracy', *New Left Review* 137–151.
63. Qiu, L. (2018) 'No, Democrats don't want "open borders"', *New York Times*, June 27.
64. Howard, T. (2010) *The Tragedy of Failure: Evaluating State Failure and Its Impact on the Spread of Refugees, Terrorism, and War* (Santa Barbara: Greenwood).
65. McKay, F.H., Thomas, S.L. and Kneebone, S. (2012) "It would be okay if they came through the proper channels': Community perceptions and attitudes toward asylum seekers in Australia', *Journal of Refugee Studies* 25(1): 113–133; McDonald, M. (2011) 'Deliberation and resecuritization: Australia, asylum-seekers and the normative limits of the Copenhagen School', *Australian Journal of Political Science* 46(2): 281–295.
66. Manne, R. (2014) 'On refugees, both the left and the right are wrong. The solution: compromise'. *The Guardian* 12 March.
67. Santhanam, L. (2018) 'Trump's immigration policy splits Americans in half, poll says', *SBS Politics*, December 11. https://www.pbs.org/newshour/politics/trumps-immigration-policy-splits-americans-in-half-poll-says
68. Grant, B., Moore, T. and Lynch, T. (eds. 2018) *The Rise of Right-Populism: Pauline Hanson's One Nation and Australian Politics* (Singapore: Springer).
69. Huysmans, J. (2006) *The Politics of Insecurity: Fear, Migration and Asylum in the EU* (London: Routledge).
70. Ngo, M. (2018) 'Germany used to be a champion of open borders in Europe. Not anymore', *Vox*, July 3. https://www.vox.com/world/2018/7/3/17530946/germany-immigration-angela-merkel-migration
71. Tajani, A. (2018) 'The migration crisis threatens to destroy the EU. We must not let it.' *The Guardian*, June 27.
72. Beeson, M. (2019) *Rethinking Global Governance* (Basingstoke: Palgrave).
73. Mounk, Y. (2018) *The People Vs. Democracy: Why Our Freedom is in Danger and how to Save it* (Cambridge, Mass.: Harvard University Press); Levitsky, S. and Ziblatt, D. (2018) *How Democracies Die: What History Reveals About Our Future* (London: Viking).
74. Inglehart, R. and Norris, P. 2016. *Trump, Brexit, and the Rise of Populism: Economic Have-Nots and Cultural Backlash*. (Harvard: Harvard Kennedy School); Rachman, G. (2019) 'Populism faces its darkest hour'. *Financial Times* January 7.
75. Significantly, 10 of the hottest years on record have occurred since 2005. Hannam, P. (2018) 'Hot start to 2019 after Australia ends its third warmest year', *Sydney Morning Herald*, December 31.

76. Interestingly, Australia's geographic isolation has generated a chronic sense of insecurity in its strategic elites and policymakers, at least; a realty that makes the adoption of progressive policies all the more difficult. See, Walker, D. (1999) *Anxious nation: Australia and the rise of Asia 1850–1939* (St. Lucia, Qld.: University of Queensland Press).
77. *The Economist* (2018) 'What the world can learn from Australia', October 27.
78. Jericho, G. (2018) 'Is it time to end the stories about the "Australian miracle"?', *The Guardian*, October 28.
79. Westcott, B. (2018) 'Australia's sixth PM in a decade. Why does it seem so ungovernable?', *CNN*, August 25. https://edition.cnn.com/2018/08/24/australia/australia-leadership-chaos-politics-intl/index.html
80. Beeson, M. and McDonald, M. (2013) 'The politics of climate change in Australia', *Australian Journal of Politics & History* 59(3): 331–348.
81. Butler, M. (2017) 'How Australia bungled climate policy to create a decade of disappointment', *The Guardian*, July 5.
82. Kelly, P. (2014) *Triumph and Demise: The Broken Promise of a Labor Generation* (Melbourne: Melbourne University Press), p. 308.
83. Rudd's successor as leader of the ALP and prime minster, Julia Gillard, was careful not to make a similar mistake, and effectively bought the support of the mining sector. Uren, D. (2010) 'Julia Gillard's $7.5bn gift to miners'. *The Australian* July 15.
84. As Guy Pearse points out, 'through their connections or presence on company boards, in neoliberal think-tanks, and through hired-gun consultants favored by the Liberal Party, Australia's biggest polluters enjoyed unparalleled access to the prime minister's office, and succeeded in having their greenhouse policy agenda adopted almost in its entirety.' Pearse, G. (2007) *High & Dry: John Howard, Climate Change and the Selling of Australia's Future* (Camberwell: Viking), p. 20.
85. Crabb, A. (2018) 'Australia's recent climate change policy: A brief history of seven killings,' *ABC*, August 23. https://www.abc.net.au/news/2018-08-23/climate-change-policy-a-brief-history-of-seven-killings/10152616
86. Elections for Australia's lower house are held every three years. Senators in the 'house of review' are elected at state level. As in the US, each state has the same number of representatives, despite very different population bases, giving a parochial bias that favors special interests. Despite the notorious inefficiencies and the shortcomings of the federal and two-party systems, the difficulty of, and willingness to, change Australia's system means it will endure despite its anachronistic shortcomings. Marsh, I. (1995) *Beyond the Two Party System: Political Representation, Economic Competitiveness and Australian Politics* (Cambridge: Cambridge University Press).

87. One of Hanson's fellow One Nation members, climate sceptic and Islamophobe, Fraser Anning, received just 19 first preference votes and yet thanks to the eccentricities of the voting system managed to end up in the Senate. McKenna, M. (2017) 'Fraser Anning is Pauline Hanson's new low-vote senator', *The Australian*, November 11.
88. Wood, D., Daley, J. and Chivers, C. (2018) 'Australia demonstrates the rise of populism is about more than economics', *Australian Economic Review* 51(3), p. 402.
89. Burchell, D. and Leigh, A. (eds. 2002) *The Prince's New Clothes: Why Do Australian's Dislike Their Politicians?* (Sydney: UNSW Press).
90. Denniss, R. (2018) 'Dead right: How Neoliberalism ate itself and what comes next', *Quarterly Essay* (70): 1–79.
91. Manne, R. (2011) 'Bad news: Murdoch's *Australian* and the shaping of the nation', *Quarterly Essay* 43, p. 113.
92. Gabbatiss, J. (2018) 'Australia refuses to ditch coal despite UN warning it must be abandoned to avert climate catastrophe'. *The Independent* 9 October.
93. Ludlow, M. and Coorey, P. (2018) 'Qld govt claims it still supports Adani', *Australian Financial Review*, February 5. There are claims that the Adani group is both corrupt and routinely flouts environmental laws in India, where it is based. Safi, M. (2018) 'Adani Group files plea in India high court to stall investigation', *The Guardian*, August 29.
94. Murphy, K. (2018) 'Climate poll shows Morrison political vulnerable as more voters back action', *The Guardian*, September 12.
95. Evans, M. and Stoker, G. (2016) 'Political participation in Australia: Contingency in the behaviour and attitudes of citizens', *Australian Journal of Political Science* 51(2): 272–287.
96. Hendriks, C. M. (2017). Citizen-led democratic reform: innovations in Indi. *Australian Journal of Political Science*, 52(4), p. 482.
97. Speaking of China, it is also worth noting that some informed observers argue that the chances of developing a form of deliberative democracy with Chinese characteristics are actually much higher than many observers (like me) seem to think. See, He, B. and Warren, M.E. (2011) 'Authoritarian deliberation: The deliberative turn in Chinese political development', *Perspectives on Politics* 9(2): 269–289.
98. AAP (2018) 'Scott Morrison tells students striking over climate change to be 'less activist', *The Guardian*, November 26.
99. Boyte, *op cit*.

CHAPTER 6

Conclusion: The Unsustainable Status Quo

Changing the way the planet works is not easy. Remarkably enough, however, human beings have managed to do just that. Unfortunately, things didn't turn out quite as planned, largely because we didn't have a plan. The popularity of the Anthropocene as the best way of describing our collective impact on the Earth tells us something important about both the state of the world and about the fact that we actually recognize our role in bringing it about. Even recognizing that we are collectively trashing—and potentially rendering unlivable—our only home is progress of a sort. While we may not have intended to make quite such a mess, we're the only ones who have the potential and the motivation to clean it up and place our habitat on a more sustainable footing, for ourselves and our fellow creatures.

Some of the possible ways of doing this are well enough known already. Given my suspicion that I am likely to be preaching to the converted given the types of people who might actually read a book like this in the first place, I shan't go into exhaustive detail about what we *ought* to do, once again. Indeed, I'm not sure we're actually obliged to do anything: as far as I'm aware, none of us actively lobbied to be here or make a small contribution to our collective problems in the first place. It's to our eternal credit, it seems to me, that any of us *do* care, not only about our own collective fate, but about the impact we're having on other apparently sentient beings. The fact that it's considered rather bad form to eat, torture or generally abuse each other these days is another of the those taken-for-granted indicators of progress that we ought to be rather pleased with ourselves about. Are the increasingly numerous vegans who live among us even more praiseworthy in that case? I think they probably are, if only because they make a more significant contribution to solving our collective problems than the rest of us do.[1]

However, in this concluding discussion I shall refrain from nagging the reader any more than absolutely necessary, not least because of my own manifold failings in this regard. Indeed, let me begin by confessing yet another of my egregious shortcomings: I am an immigrant of precisely the sort that I have spent some time suggesting may be part of our collective problems. Some of my arguments may consequently seem at best hypocritical, at worst, selfish and even heartless. Pulling up the drawbridge after reaching the sanctuary of what has turned out to be a rather commodious castle, replete with the world's largest moat, doesn't display the generosity of spirit one might hope to see from someone urging collective action to avoid a common fate. At the risk of laboring the point, however, a central part of the—admittedly rather instrumental and self-serving—argument of this book is that we need to act where we live first. If the political circumstances we inhabit at the local level aren't conducive to sustaining 'progressive' politics, then we can't begin to take the sorts of meaningful action that may ultimately have benefits for people in other parts of the planet, too. Environmental populism needs to be popular.

There is, of course, a continuing debate about what progressive politics might look like and whether the conditions that might sustain them have *already* disappeared. Is the best 'we'—in this case the fortunate inhabitants of places such as Australia—can hope for just insulating ourselves from the problems of the rest of the world for as long as possible? In this context let me make another confession: as far as I'm concerned, I'm pretty confident that life will remain bearable, possibly even quite agreeable for the remainder of my existence.[2] In the interests of full disclosure, I should also point out that I'm childless, so I don't share the same sort of immediate concerns that, I assume, parents feel about the future welfare of their offspring. I say 'assume' because plainly either many people don't especially care about the possible future their children will inherit, or they remain in furious denial despite the overwhelming evidence to the contrary.

The reader won't be surprised to learn by this stage that there is an extensive and erudite literature on the importance of 'intergenerational justice', and the possibility that as yet unborn generations have rights, and we consequently have an obligation not to leave the place in too much of a mess.³ Some hopeful souls have argued that 'it is likely that this movement of thought will increasingly percolate into popular as well as official consciousness.'⁴ Good luck with that. Clearly, no one has told Donald Trump. More fundamentally, I'm not sure we've got enough time left to wait for the percolation process to take effect. Direct political action and ramping up the pressure on those individuals who have chosen to occupy leadership positions look more likely to have an impact.

WHAT IS TO BE DONE?

One of the key conclusions that I hope emerges from this book is about the perennial value of thinking globally and acting locally. We need to be aware of the nature of the problems we confront—which means accepting the best guesses and judgments of recognized authorities on climate change about what's happening to the environment, and about what might be done to rectify it—and then acting where and how we can. Peter Katzenstein, one of the most influential political scientists of his generation, argues that we are witnessing the emergence of a new form of 'protean power', which is the 'effect of improvisational and innovative responses to uncertainty that arise from actors' creativity and agility in response to uncertainty.'⁵ The unexpected, unplanned and locally realized fall of the Berlin Wall is the most significant example of this phenomenon, in which contingent circumstances present unexpected opportunities for action and change. Yet even if the reader accepts the argument that the best hope we collectively have for generating meaningful action is by placing populist-style pressures on 'our' politicians, policymakers and plutocrats—our very own ruling classes, if you will—the precise nature of our political engagement will reflect our own distinctive circumstances and the national political framework within which it is embedded. The chances of acting, and the possible costs of doing so, are likely to vary profoundly as a consequence, too.

As we have seen, despite the great hopes that were held about the prospects for a global democratic revolution generally and the Arab Spring in particular, actually creating the conditions in which progressive politics can flourish remains a formidably difficult challenge. The remarkable thing is that we have democracies everywhere, perhaps; we take them for granted and forget how fragile they are at our collective peril. The point about democratic regimes is that they provide potential points of access to the political process that are either not available or are closely guarded by privileged insiders in other systems. True, many democracies are now dominated, captured and corrupted by self-serving, cronyistic elites who structure the political system to reflect and further their own interests. But the fact remains that engagement is not only generally possible, but the simple act of participating may have transformative effects, not least on the participants themselves. As Mounk points out, 'the stability of the [democratic] system may thus depend on finding ways to make ordinary citizens feel that they are in charge again.'[6]

In nondemocratic, authoritarian regimes, taking on the ruling elites—not to mention the military, whose loyalty is often secured through patronage politics[7]—is a dangerous and often bloody business.[8] Even with subsequent economic development, there is no guarantee that democratic rule will be embraced or sustained. Expecting the populations of such countries to lead the movement for global collective action on climate change or anything else when their immediate concerns may be more to do with questions of survival looks like wishful thinking. The rather unfashionable figure of the late Samuel Huntington may not have been right about everything, but he did have a point about the importance of functioning political systems with the ability to act: 'Without strong political institutions, society lacks the means to define and to realize its common interests. The capacity to create political institutions is the capacity to create public interests.'[9]

Those of us fortunate enough to live in countries where such institutions already exist, therefore, have not just an opportunity, but perhaps even an obligation, to act. As Amartya Sen observed, 'as competent human beings, we cannot shirk the task of judging how things are and what needs to be done…It is not so much a matter of having exact rules about how precisely we ought to behave, as of recognizing the relevance of our shared humanity in making the choices we face.'[10] There is extensive literature on the importance and ethical consequences of economic inequality, too.[11] But even those who are intellectually or emotionally unmoved by such arguments might concede that the existence of growing inequality within countries, and the failure of much of the rest of the world to 'catch up', presents a major challenge for domestic social stability and international security.

Yet even if we can agree that something ought to be done, the question is what exactly. More pointedly, does it necessitate a complete transformation of the economic, political and social order, as some claim? Ultimately, it might, and populist upsurges from below do seem to fit the criteria that prominent theorists argue are essential to 'social revolutions' of a sort that transform existing societal and political structures in a mutually reinforcing fashion.[12] No doubt we become more conservative as we age—and that may indeed be part of the collective difficulty we face given the political dominance of the gerontocracy[13]—but it's worth remembering that it's a lot easier to tear things down than it is to build things up. As Mao famously observed, 'a revolution in not a dinner party'. He would know; the history of the Cultural Revolution is a painful and instructive reminder that encouraging young people to take over doesn't always end well either.[14]

For all its manifold shortcomings and contemporary problems, as Churchill famously observed, there are many worse political systems than democracy. Democracy may be an inherently nonrevolutionary system,[15] but that means that those of us who are fortunate enough to live in democratic systems, even poorly performing ones, may have to work with what we've got in the short term. Such an outcome may not actually be the end of the world though, so to speak, and looks more politically feasible and potentially popular. The trick is to make democracy work more effectively and encourage elected representatives to treat climate change with the seriousness it merits. Electing people with progressive ideas who actually understand the significance of climate change and who are equally determined to do something about it is one place to start.[16] But as the brief discussion of Australia's recent political history in Chap. 5 demonstrated, even electing politicians who understand the science and who are not diehard apologists for the coal industry is no guarantee of change. Nevertheless, Australia's political class really *is* potentially well placed to play a useful and constructive role at home and abroad. I shall conclude by explaining why Australia could provide an improbable but not impossible role model for the world.

Australia Saves the World…

Let me reemphasize that I don't think Australia or the people who live there are especially meritorious, deserving of universal admiration, or of fulfilling a unique historical destiny. On the contrary, it's possible to make a pretty good case for saying that Australians and their political class have often been slightly paranoid, insular and complacent, and remarkably uninterested in what's going on in the rest of the world. Likewise, the less said about the treatment of the indigenous population and Australia's unique wildlife the better, perhaps.[17] Nevertheless, by world standards it's an oasis of privilege, prosperity and peace. With all its natural advantages, in fact, even mediocre political leaders with next to no sense of purpose or vision could run the place. And as Donald Horne pointed out, that's generally just what has happened.

At a time when Australia is at the sharp end of global warming, though, such vacuous insularity has become not simply morally indefensible, but unsustainable in every sense of the word. If Australia's leaders are serious about being a responsible 'middle power' and a good international citizen,[18] they really need to start acting the part. Far from championing the continuing domestic use and export of coal, for example, Australia's political leaders should immediately stop the development of new mines, while simultaneously moving to close down existing operations and the use of greenhouse gas emitting coal to produce energy. Given the amount of free energy available from the sun, one would have thought this was the proverbial no-brainer in Australia, unless one also considered the influence of the coal lobby and the possible importance of mining jobs in marginal electorates.[19]

Even nonrevolutionary change is likely to produce a few casualties, but such is life, as Ned Kelly memorably observed.[20] Crucially, very few people are actually directly employed in the mining industry,[21] and Australia is wealthy enough to compensate and/or retrain those who lose their jobs as a consequence of the economic restructuring that is an inescapable and urgent part of Australia's (and the world's) response to climate change. Indeed, 'Australia' could be significantly wealthier if its leaders chose not to invest in a range of unproven, ultimately ineffective new weapons systems and spend the money on revamping the economy instead. It is, however, testimony to the deadening impact of the conventional wisdom, especially its impoverished understanding of security, that the defense establishment received $150 billion for a new submarine fleet and another $17 billion for a new generation of (trouble-plagued) aircraft with next to no public debate.[22]

This is not the place for a detailed discussion of Australia's 'strategic posture'.[23] But the continuing willingness of successive governments to pour money into things that will almost certainly never be used in anger, and which would make no decisive difference to the outcome of any great power war even if they were, is remarkable. It is also depressingly and profoundly at odds with the very real, increasingly urgent danger that Australia actually *does* face. Even half of the proposed $150 billion spending on submarines could help retro-fit national infrastructure along more sustainable lines, compensate the immediate 'losers' of economic restructuring, and possibly even offer aid and development assistance to those countries that are currently contributing to the international migration crisis. All of which would be of greater value than submarines that will very likely be outdated and redundant before they actually go into service.[24]

Other countries have different domestic problems to overcome, but we all have—or ought to have—the same priority: without a sustainable environment and a habitable planet, we all face a nightmarish future from which bigger bombs can't protect us. It is unlikely in the extreme that 'great powers' such as China and the US, who are currently locked in an increasingly dangerous competition for strategic dominance and influence in the Asia-Pacific, will significantly change their current security policies.[25] Given that their own security establishments increasingly recognize that climate change is a major security threat, this is more surprising than it might seem. It is not possible or useful to launch into an exhaustive discussion of the 'security dilemma' or the futility of arms races,[26] to recognize that possible leadership from the US and China is compromised by their rivalry.[27]

In such a context, it is possible that enlightened middle powers, which are not necessarily or inevitably part of this 'structural' rivalry, could—in theory at least—act in more creative, constructive and even inspiring ways. Australia could undoubtedly 'get away with' having a vastly reduced defense budget, in just the same way as its smaller neighbor New Zealand has.[28] It could also demonstrate to some of its neighbors that alternative policy priorities are not only possible, but that they also contribute to more sustainable, productive domestic development. No doubt this will seem like wishful thinking, possibly even irresponsible to some readers. After all, the first priority of policymakers is to preserve the security of the nation. Quite so. But the sorts of security threats faced by Australia are now coming from a rapidly deteriorating environment, *not* from the threat of Chinese invasion. Indeed, in the event of our collectively failing to solve the climate change challenge it is entirely possible that China—and many other countries—will become more aggressive and authoritarian, and even more of a threat than it currently is.[29]

Hope Springs Eternal

Large-scale, incessant, focused pressure from below—a populist upsurge, if you like—*could* make a difference, and possibly not just in liberal democracies such as Australia. But that's where I live and can act, if only by not voting for people who fail to make addressing climate change their greatest priority. People in other countries will have to work out how they can act within the confines of what are too frequently less forgiving and inclusive political systems. The risks are altogether more serious and consequential, and that is yet another reason why those who can act should; and why those who could provide leadership really ought to do so—or get out of the way of those that might. I am not convinced that people of my generation have the capacity or the motivation to do so, however.

The period we grow up in profoundly shapes our values and expectations of what is possible. The generation that grew up during the Great Depression was famously often frugal and alert to the possibility that things could fall apart disastrously at any moment. Not a bad orientation to reality in the present day as it happens. Uplifting as Steven Pinker's books are, for example, he retains his generation's optimism about our ability to come up with technical fixes for even the most intractable and complex of problems.[30] Perhaps geoengineering or some other technological fix will turn up and we'll all live happily ever after. You wouldn't want to bet on it though, and the precautionary principle still looks like one we need to take seriously.[31]

Young people are the great hope, not least because they have grown up with a sense of looming crisis. They 'get' climate change in a way their elders, even well-intentioned ones, generally don't. Things are unlikely to get that bad before I expire, but they could get really *really* bad by the end of the century when some of today's youngsters could actually still be around. Their offspring—if they're game to have any in such circumstances—will almost certainly have to confront unspeakable horrors if we don't get our collective act together very quickly.

Yet hope springs from the most unlikely sources at times. Easily the most inspiring speech of the annual meeting of the world's overprivileged, overconsuming, overrated movers and shakers at the Davos World Economic Forum of 2019 was delivered by 16-year-old Greta Thunberg (who is almost inevitably from civilized Sweden). Greta's message was compelling, clear and worth quoting:

> Either we prevent temperatures from rising above 1.5 degrees (Celsius), or we don't. Either we avoid chain reaction of unravelling ecosystems, or we don't. That's as black or white as it gets. Now we all have a choice: we can either create transformational action or continue with business as usual and fail.[32]

Greta wants the older generation to 'panic' about the implications of runaway climate change in the same way she is doing. I think my panicking days are behind me, Greta—other than about missing appointments and forgetting people's names—but I take your point. Unfortunately, I'm not sure the Davos crowd will. Something like 1500 private jets flew to Switzerland to discuss climate change and the merits of 'globalization 4.0'—whatever that is.[33] Bridging the disjuncture between the way a tiny global elite and the rest of us live, even in privileged places like Australia, is not going to be easy. But as the Occupy movement usefully pointed out, there are only 1 percent of them and 99 percent of us. Acting to redress the balance of opportunity and influence ought to be popular, at least.

Good luck to you and your generation, Greta. I fear you'll need it—unless a global upsurge of environmental populism saves the day, of course.

Notes

1. Unfortunately, even this is not as straightforward as it might seem. On the plus side, a lot of animals won't die in unpleasant conditions. See Henderson, E. (2018) 'Why veganism isn't as environmentally friendly as you might think', *The Independent*, January 27.
2. At the time of writing, I'm 66 and don't expect (or especially want) to live more than another 20 years or so at the outside.
3. Hiskes, R.P. (2009) *The Human Right to a Green Future: Environmental Rights and Intergenerational Justice* (Cambridge: Cambridge University Press).
4. Holden, B. (2002) *Democracy and Global Warming* (London: Continuum), p. 80.
5. Seybert, L.A. and Katzenstein, P.J. (2018) 'Protean power and control power: Conceptual analysis', in, L.A. Seybert and P.J. Katzenstein *Protean Power: Exploring the Uncertain and Unexpected in World Politics* (Cambridge: Cambridge University Press), p. 4.
6. Mounk, Y. (2018). 'The undemocratic dilemma', *Journal of Democracy*, 29(2), 98–112.
7. Beeson, M. and Bellamy, A.J. (2008) *Securing Southeast Asia: The Politics of Security Sector Reform* (London: Routledge).
8. At the time of writing (early 2019), the militaries of Zimbabwe and Venezuela were providing instructive reminders of the potentially destructive role of military establishments whose principal focus is domestic and self-serving.
9. Huntington, S.P. (1968) *Political Order in Changing Societies* (New Haven: Yale University Press), p. 24.

10. Sen, A. (1999) *Development as Freedom* (Oxford: Oxford University Press), p. 283.
11. See, for example, Pogge, T.W. (2008) *World Poverty and Human Rights* (Cambridge: Polity).
12. Skocpol, T. (1979) *States and Social revolutions: A Comparative Analysis of France, Russia and China* (Cambridge: Cambridge University Press), pp. 4–5.
13. Fox, J. (2018) 'The long baby boomer reign isn't ending just yet', *Bloomberg*, April 29; Beeson, M. (2019) 'Green New Deal: The rebirth of American leadership? *Australian Outlook*, February 6. http://www.internationalaffairs.org.au/australianoutlook/green-new-deal-us-politics/
14. Clark, P. (2008) *The Chinese Cultural Revolution: A History* (Cambridge: Cambridge University Press).
15. Motyl, A.J. (1999) *Revolutions, Nations, Empires: Conceptual Limits and Theoretical Possibilities* (New York: Columbia University Press), p. 34.
16. There are some hopeful signs. Remnick, D. (2018) 'Alexandria Ocasio-Cortez's historic win and the future of the Democratic Party', *The New Yorker*, July 23.
17. But if you really must know, Cox, L. (2018) '"A national disgrace": Australia's extinction crisis is unfolding in plain sight', *The Guardian*, February 13; Conifer, D. (2017) 'Closing the gap: Australia is failing on Indigenous disadvantage goals', *ABC*, February 14. https://www.abc.net.au/news/2017-02-14/closing-the-gap-report-card-failing/8268450
18. Beeson, M. and Higgott, R. (2014) 'The changing architecture of politics in the Asia-Pacific: Australia's middle power moment?', *International Relations of the Asia-Pacific* 14(2): 215–237.
19. Karp, P. and Evershed, N. (2018) 'Coal lobby ads biggest third-party political expenditure in Australia', *The Guardian*, February 1.
20. Ned Kelly is Australia's most notorious Bushranger or outlaw. Legend has it these were his final words before he was hung.
21. Mining as a whole employs less that 2 percent of the population in Australia. A similar picture exits in the US. See, Vandenbroek, P. (2018) *Employment by industry statistics: a quick guide* (Canberra: Parliament of Australia); DiChristopher, T. and Schoen, J. (2018) 'Trump says 'the coal industry is back.' The government's jobs numbers say otherwise', *CNBC*, August 23.
22. Wroe, D. (2016) 'Defence white paper: new submarine fleet to cost taxpayers $150 billion', *Sydney Morning Herald*, February 25; Seldel, J. and Bickers, C. (2017) 'Australia's F-35A stealth fighters may cost millions to bring up to a fighting standard', *news.com.au*, https://www.news.com.au/technology/innovation/australias-f35a-stealth-fighters-may-cost-millions-to-bring-up-to-a-fighting-standard/news-story/7a28b881ff92cc27

ed5fd88f2480f512; Gottliebsen, R. (2018) 'Defence should be first target of national integrity commission'. *The Australian* December 17.
23. Hugh White is one of the more thoughtful mainstream critics. See, White, H. (2016) 'Australia's Defence White Paper has a hidebound view of Asia's future'. *East Asia Forum* 8 March; White, H. (2017) 'Without America: Australia in the new Asia', *Quarterly Essay* 681–81.
24. Gottliebsen, R. (2017) 'Defence submarine contract analysis reveals another costly mistake'. *The Australian* September 27.
25. Sanger, D.E. and Broad, W.J. (2019) 'U.S. suspends nuclear arms control treaty with Russia'. *New York Times* February 1.
26. See, Booth, K. and Wheeler, N.J. (2008) *The Security Dilemma: Fear, Cooperation, and Trust in World Politics* (Basingstoke: Palgrave).
27. Beeson, M. (2017) 'Why has leadership in the Asia–Pacific proved so elusive?', *Chinese Political Science Review* 2(4): 567–581.
28. Beeson, M. (2015) 'Australia's defence: should we go down the Kiwi road?', *The Strategist*, December 10, http://www.aspistrategist.org.au/australias-defence-should-we-go-down-the-kiwi-road/
29. Beeson, M. (2010) 'The coming of environmental authoritarianism', *Environmental Politics* 19(2): 276–294.
30. Pinker, S. (2018) *Enlightenment Now: The Case for Reason, Science, Humanism, and Progress* (New York: Viking), p. 134.
31. O'Riordan, T. (2013) *Interpreting the Precautionary Principle* (London: Routledge).
32. Workman, J. (2019) '"Our house is on fire." 16 year-old Greta Thunberg wants action', Global Agenda, *World Economic Forum*, January 25. https://www.weforum.org/agenda/2019/01/our-house-is-on-fire-16-year-old-greta-thunberg-speaks-truth-to-power
33. Ratcliffe, R. (2019) 'Record private jet flights into Davos as leaders arrive for climate talk', *The Guardian*, January 23.

Index[1]

A
Achen, Christopher, 71
Adani, 102
Africa, 63
Alternative for Germany (AFD), 95
Anthropocene, 9, 111
   unprecedented nature of, xiii
Arab Spring, 75, 94, 95
Asia, 63
Aslanidis, Paris, 69
Australia, xiv, 10, 11, 14, 27–29, 48, 51, 52, 96–98, 100–103, 109n75, 109n76, 109n84, 109n86, 112, 115–117, 119, 120n21
   business in, 28, 102
   China and, 117
   coal and, 51, 102
   defense policy, 117
   elites in, 100
   environmental policies of, 101, 117
   environment of, 14, 51, 101
   leadership of, 102
   lucky country and, 100
   as a middle power, 116
   migration and, 97
   national interest and, 51, 100
   politics and, 10, 51, 101, 115
   as a role model, 115
Australian Labor Party (ALP), 101
Authoritarianism, 76, 114
   in China, 30
   environment and, 44
   populism and, 69

B
Bank of England, 71
Barbier, Edward, 33
Bartels, Larry, 71
Belt and Road Initiative (BRI), 31
Berlusconi, Silvio, 68
Biosphere, 66
   degradation of, 11–13

---

[1] Note: Page numbers followed by 'n' refer to notes.

Bonaparte, Napoleon, 44
Boyte, Harry, 95, 103
Brexit, 71
Britain/UK, 46, 74
   migration and, 73
Brown, Mark, 7
Brzezinski, Zbigniew, 42
Buchanan, Pat, 91
Bush, George W., 43
Business interests
   in Australia, 102
   lobbying by, 28, 92
   regulation and, 33, 92
   structural power of, 28, 49, 102

C

Cambodia, 31
Canovan, Margaret, 71
Capitalism, 21
   in Australia, 29
   Cold War and, 65
   democracy and, 43
   environmental impact of, 24–28
   future of, 23, 65, 66
   global scale of, 24
   models of, 28–32
   in the US, 29
Carbon sinks, 9
Carbon tax, 34
Center for New American Security (CNAS), 47
Center for Strategic and International Studies (CSIS), 47
Chavez, Hugo, 67
Chernobyl, 25
China (People's Republic of/PRC), 10, 22, 23
   Australia and, 119
   authoritarianism and, 30, 53, 91
   crises and, 91
   decentralisation in, 52–53
   elites in, 30
   environmental impact of, 32, 35, 48
   environmental policies of, 28, 91
   foreign policy of, 30
   as a great powers, 117
   history of, 89
   leadership in, 13, 30, 32
   politics in, 10, 75
   social media in, 75
   United States and, 47, 48, 53–54, 117
Chinese Communist Party (CCP), 29, 30, 32
   social control and, 91
Christoff, Peter, 50
Churchill, Winston, xiii, 115
Civilization, 10–14
   capitalism and, 25
   economic development and, 21
Civil society, 66
   global, 75–79
Clapp, Jennifer, 34
Climate change, 8, 9
   collective action and, xiv, 112, 114
   conflict and, 4
   economic challenge of, xiv
   global governance and, 49
   impact of, 4, 22
   as realistic threat, 43, 48
   science and, 1, 3–4, 7–10
   social unrest and, 44
   speed of, 12
   as a wicked problem, 3, 12
   See also Global warming
Climate denial, 52
Coal industry, 9
   attitudes of, 27
   in Australia, 51, 102, 115

EPA and, 27
lobbying by, 27, 28, 101
Cold War, 65
Colgan, Jeff, 52
Collective action
  climate change and, xiv, 114
  democracy and, xv
  obstacles to, xiv, 2, 21–22, 35, 42, 49, 73, 78–80, 92, 99, 114
  states and, 2
  transnational, 78
  urgency of, 35
Collier, Paul, 96
Common but differentiated responsibilities, 50
Concert of Europe, 44
Conservation, 26
  Polanyi and, 76
Consumption, 29
  unsustainable, 25, 26, 32, 75
Cooperation, xiv, 2, 13, 23, 30, 32, 34, 41–54, 99
Copernicus, Nicolaus, 6
Corbyn, Jeremy, 63, 74
Cosmopolitanism, 67, 97
  democracy and, 77
Critical junctures, 34
Cultural Revolution, 115
Cultural values, 71

D
Dauvergne, Peter, 26, 34
Davos World Economic Forum, 118
Dawkins, Richard, 9
Decentralization, 52, 53
Degrowth, 32
Democracy, 47
  capitalism and, 43
  climate change and, xv
  cosmopolitan, 77
  death of, 64–67
  degrowth and, 33
  geopolitics and, 69
  migration and, 98
  monitory, 66
  nonrevolutionary nature of, 115
  pathologies of, 28
  populism and, 75
  Third Wave of, 65
  in the US, 28–29
  value of, 113
  young people and, 64, 73
Demography, 13
  China and, 29
  lifestyles and, 25
  populism and, 71
  poverty and, 22
Deng, Xiaoping, 35
Diamond, Jared, 14
Drezner, Daniel, 50
Dryzek, John, 52
Dzur, Albert, 79

E
Eckersley, Robyn, 50
Ecological hegemony, 29
Ecological rift, 29
Economic development, 11, 13, 21–35, 65, 93, 114
  in China, 22
  in East Asia, 22
  inequality and, 23
  models of, 2
  political change and, 65
  prospects for, 41, 113
  sustainability and, 5, 13
  working class and, 72
Economic theory, 26, 27
Ecosystems services, 26
Eco-tax, 64

Einstein, Albert, 5
Elites, 97
  attitudes of, 4, 71
  in Australia, 100
  authoritarian, 114
  in China, 30
  corruption of, 68, 69, 114
  cosmopolitan, 73, 91, 97
  democracy and, 71
  economic, 27, 33, 34, 91
  expertise of, 7, 71
  failures of, 66, 87, 97
  inequality and, 73
  liberal, 70
  local, 53
  military, 4, 80
  political, 27, 43, 103
  populism and, 69
  ruling, 6
  survival strategies of, 48
  transnational, 94
  Trump and, 68
  in the US, 68, 70
Enlightenment, 6, 102
Environment
  collective goods and, 2
  crises and, 9, 14, 53
  humanity and, 11, 12, 21
  multilateralism and, 50
  neoliberalism and, 75
  public policy and, 5
  regimes and, 10, 51
  as a security threat, 45, 46, 48, 80, 117
Environmental Protection Agency (EPA)
  coal industry and, 27
Ethics, 10
European Union (EU)
  austerity and, 95
  importance of, xiv, 45
  migration and, 99
  populism and, 41
  sovereignty and, 45
Expertise, 4
  elites and, 71
  loss of confidence in, 7–8, 89

F

Facts, 1–3, 7–10
Fake news, 3
Fischer, Frank, 67
Foster, John B., 29
France, 44
Fraser, Nancy, 94
Fukuyama, Francis, 64, 65, 73, 74

G

G20, 34
Gaia hypothesis, 9
Galileo, Galilei, 6
Garnaut, Ross, 92
Geo-engineering, 14, 118
Geopolitics, 42
  democracy and, 69
Gerbaudo, Paolo, 94
Germany, 52
  immigration and, 70, 98
  populism and, 88
Gilets jaunes (yellow vests), 63, 64, 93
Global financial crisis (GFC), 48, 90, 91, 93
Global governance, 5
  accountability and, 45
  climate change and, 49
Globalization, 5, 46
  acting locally and, 113
  impact on traditional societies, 72
Global warming, 8, 9
  unprecedented nature of, 3
  See also Climate change
Golden Dawn, 95

Gore, Al, 49
Government-organized NGOs (GONGO), 76
Gramsci, Antonio, 80
Grand strategy, 43
Great acceleration, 11
Great Depression, 90
Greece, 94, 95
Green, Jessica, 52
Greenhouse effect, 4, 9
Greenhouses gases, 8–11
Green New Deal, 33
Greens, 102

H
Habermas, Jürgen, 66
Hacker, Jacob, 28
Hamilton, Clive, 14
Hanson, Pauline, 98
Hausknost, Daniel, 32
Hendricks, Carolyn, 79, 102
Hitler, Adolf, 88, 90
Horne, Donald, 100, 116
Human nature, 2
Human rights, 97, 98
Huntington, Samuel, 65, 114

I
Identity politics, 72–74
Ideology, 2, 6, 95
  Cold War and, 65
Individualism, 2, 21
  politics and, 72–74, 95
Inequality, xiv, 114
  development and, 21, 23, 25
  social divisions and, 73
  the US and, 89, 90
Inglehart, Ronald, 71
Intergenerational justice, 2, 21, 103, 112, 119n3

Intergovernmental organizations (IGOs), 50
Intergovernmental Panel on Climate Change (IPCC), 11, 12, 50, 102
  Trump administration and, 28
International community, 46
International financial institutions (IFIs), 24
International relations (IR), 2
  critical, 43, 75
  institutions in, 45
  realism and, 46–48
Islam, 78

J
Japan
  environmental impact of, 24
Judi, John, 91

K
Kaltwasser, Cristóbal, 69
Katzenstein, Peter, 113
Kauffmann, Eric, 73, 97
Kazin, Michael, 89
Keane, John, 66
Keck, Margaret, 77
Kelly, Ned, 116
Kelly, Paul, 101
Keynes, John Maynard, 89, 90
Kissinger, Henry, 42
Klein, Naomi, 28
Kovel, Joel, 25
Krasner, Stephen, 45
Kuttner, Robert, 46

L
Laclau, Ernesto, 75, 93
Lasch, Christopher, 72
Latin America, 63

Leadership, xiii
  in Australia, 102
  charismatic, 71
  in China, 13, 30, 32
  failures of, 13, 51
  need for, 51, 52
  states and, 50
  Trump and, 54
Learning
  public policy and, 5
Liberalism, 69
Liberal-National Party coalition, 102
Lo, Kevin, 53
Lobbying, 28, 30
Lomborg, Bjørn, 91
Lovelock, James, 9

M
Macron, Emanuel, 63
Manne, Robert, 102
Mao, Zedong, 25, 35, 87–89, 115
Marxism, xiv, 69
  capitalism and, 25
  ecological rift and, 29
Mearsheimer, John, 43
Mekong River, 31
Merkel, Angela, 70, 99
Mexico, 98
Middle East, 48, 95
Middle powers, 116, 117
Migration
  Australia and, 98
  crisis of, 95–100
  environment and, 96
  Germany and, 70
  Sweden and, 70
  UK and, 73
  US and, 97
Modernization, 74
Mouffe, Chantal, 75, 93
Mounk, Yascha, 114

Movement of the squares, 94
Mudde, Cas, 68, 69

N
National interest, 27, 45
  Australia and, 51
  contingent nature of, 43
  privileging of, 44
  security and, 44
Nazi Germany, 69, 88
Neoliberalism, 24
  critiques of, 91
  environment and, 75
New Deal, 90, 92
Newell, Peter, 33
Newton, Isaac, 5
New Zealand, 49, 117
Nongovernmental organizations
    (NGOs), 76
  environmental, 53
Norris, Pippa, 71
North, the
  global, 45
  imperial mode of living in, 29
  zones of peace and conflict and, 48
Nuclear war, 3–4

O
Obama, Barak, 49
Objectivity, 7
Occupy movement, 91, 93, 119
Oil, 9
Olson, Mancur, 2
Orban, Viktor, 67
Ostrom, Elinor, 2

P
Paris Agreement, 47, 51
Path-dependency, 23

People's Party, 89
Perón, Eva, 67
Perón, Juan, 67
Perrot, Ross, 91
Pierson, Paul, 28
Piketty, Thomas, 90
Pinker, Stephen, 118
Podemos, 93
Polanyi, Karl, 76
Policy entrepreneurs, 92
Politics
  in Australia, 10, 11
  in China, 10
  decline of traditional forms, 72
  Left, 73–75, 91
  mobilization and, 10, 75, 77, 103
  progressive, xiii, 10, 64, 73, 90, 92, 96, 103, 113, 115
  racial, 90, 96
  rigging of, 67
  Right, 73, 75
  in the United States, 11, 28, 33, 48, 64, 70, 89–93
  value of participation, 103, 114
Populism, xiii, xv, 7, 13, 28, 41, 43, 48, 63–80, 87–103, 112, 119
  appeal of, 67, 70, 71
  authoritarian, 13
  bad, 43, 87–90, 95
  contradictions of, 93–95
  elites and, 69
  environmental, 112, 119
  everyday politics and, 95
  the people and, 69–72, 95
  political economy of, 90–93
  progressive politics and, xiii, 74–76, 89, 119
  reputation of, xiii
  theories of, 63, 67–74
  transnational, 77
Poverty, 22
  development and, 27

Progress
  environmental, 26
  indicators of, 112
  possibility of, 7–8, 41

R
Rao, D.N., 31
Reagan, Ronald, 24
Realism, 42–44, 65, 69
  states and, 44
Regulation, 92
  business interests and, 33
Religious belief, 6, 7
Renewable energy
  in China, 30
Revolution, 33, 115
  Arab Spring and, 75
  leaderless, 93
Rothkopf, David, 92
Rudd, Kevin, 51, 101

S
Sackett, Penny, 12
Sandel, Michael, 70
Sanders, Bernie, 74
Schreurs, Miranda, 53
Science, 102
  climate change and, 1, 3–5
  contestation of, 5
  development of, 5–7
  natural *vs.* social, 4–5
Security
  dilemma, 117
  environment and, 45, 46, 48, 117
Sen, Amartya, 114
Shapiro, Ian, 67
Shareholder activism, 27
Shue, Henry, 52
Sikkink, Kathryn, 77
Smith, Neil, 23

Social capital, 72
Social construction of reality, 4–8
Social media, 75
  Occupy Wall Street and, 93
Social movements, 76–79
South, the, 23
  environmental problems in, 23, 29
Sovereignty
  imperial, 76
  pooling of, 45
  popular, 95
Soviet Union, 25, 46, 65
  environmental impact of, 35
Spain, 93, 94
Spash, Clive, 27
Stalin, Joseph, 87
States
  collective action and, 1
  conflict and, 4
  continuing importance of, 50
  international expansion of, 44
  liberalism and, 43
  possible role of, 44–46, 91
Steffen, Will, 12, 54
Stevenson, Hayley, 52
Subjectivity, 7–8
Sweden
  immigration and, 70

T
Taggart, Paul, 69
Tarrow, Sidney, 77
Tea Party, 91
Technology, 14
Thatcher, Margaret, 24, 80
Thirty Years' War, 44
Thunberg, Greta, 118
Tilly, Charles, 77
Trade unions, 74
Transnational public sphere, 94
Treaty of Westphalia, 44

Trump, Donald, 30, 113
  climate denial and, 52
  elites and, 68
  foreign policy and, 68
  leadership and, 54
  migration and, 98
  multilateralism and, 47
  populism and, 43, 63, 68–70, 88
Trump administration
  attitude to fossil fuels, 28
  Paris Agreement and, 51
Turnbull, Malcolm, 51, 102

U
Unequal ecological exchange, 23
UN Framework Convention on
    Climate Change (UNFCCC), 50,
    52
United Nations Environmental
    Programme (UNEP), 33
United Nations (UN), 45
United States (US)
  business in, 27, 28, 30
  China and, 47, 48, 53, 117
  decentralisation in, 52
  democracy in, 28, 68, 69
  as a great powers, 117
  hegemony and, 46, 54
  inequality in, 90
  migration and, 97
  politics in, 11, 28, 33, 48, 64, 68,
    70, 99
  populism and, 73, 89, 95
  post-war international order and, 43
UN's World Congress of Local
    Governments for a Sustainable
    Future, 52

V
Vietnam, 31

W
Weber, Max, 5
Western Europe, 44, 65
  migration and, 98
  populism and, 73
West, the, 6
  modernity and, 72
  success of, 13, 74

Wheeler, Andrew, 28
World Bank, 93
World War I, 89
World War II, 11, 43, 46, 76

X
Xi, Jinping, 32, 88

GPSR Compliance

The European Union's (EU) General Product Safety Regulation (GPSR) is a set of rules that requires consumer products to be safe and our obligations to ensure this.

If you have any concerns about our products, you can contact us on

ProductSafety@springernature.com

In case Publisher is established outside the EU, the EU authorized representative is:

Springer Nature Customer Service Center GmbH
Europaplatz 3
69115 Heidelberg, Germany

www.ingramcontent.com/pod-product-compliance
Ingram Content Group UK Ltd.
Pitfield, Milton Keynes, MK11 3LW, UK
UKHW041414180426
11947UKWH00007B/132